Student Support Materials for

AQA

D1358649

A2 CHEMISTRY

Module 5: Thermodynamics and Further Inorganic Chemistry

Geoffrey Hallas
Andrew Maczek
David Nicholls

This booklet has been designed to support the AQA Chemistry A2 speci cation. I t contains some material which has been added in order to clarify the speci cation. The examination will be limited to material set out in the speci cation document.

Published by HarperCollins*Publishers* Limited
77 85 Fulham Palace Road
Hammersmith
London W6 8JB

| www.**Collins**Education.com |
| Online support for schools and colleges |

© HarperCollins*Publishers* Limited 2001
First published 2001
Reprinted 2001 (twice)

ISBN 0 00 327705 4

Geoffrey Hallas, Andrew Maczek and David Nicholls assert their moral right to be identi ed as the authors of this work.

All rights reserved. No part of this publication may be reproduced, stored in a retrieval system, or transmitted in any form or by any means, electronic, mechanical, photocopying, recording or otherwise, without either the prior permission of the Publisher or a licence permitting restricted copying in the United Kingdom issued by the Copyright Licensing Agency Ltd., 90 Tottenham Court Road, London W1P 0LP.

British Library Cataloguing in Publication Data
A catalogue record for this publication is available from the British Library

Writing team: John Bentham, Colin Chambers, Graham Curtis, Geoffrey Hallas, Andrew Maczek, David Nicholls
Front cover designed by Chi Leung
Design by Barking Dog Art, Gloucestershire
Printed and bound by Scotprint Ltd, Haddington

The publisher wishes to thank the Assessment and Quali cations Alliance for permission to reproduce the examination questions.

| You might also like to visit |
| www.**fire**and**water**.com |
| The book lover s website |

Other useful texts

Full colour textbooks
Collins Advanced Modular Sciences: Chemistry A2
Collins Advanced Science: Chemistry

Student Support Booklet for Chemistry A2
AQA Chemistry Module 4: Further Physical and Organic Chemistry

What books do I need to study this course?

You will probably use a range of resources during your course. Some will be produced by the centre where you are studying, some by a commercial publisher and others may be borrowed from libraries or study centres. Different resources have different uses – but remember, owning a book is not enough – it must be *used*.

What does this booklet cover?

This *Student Support Booklet* covers the content you need to know and understand to pass the module test for AQA Chemistry A2 *Module 5: Thermodynamics and Further Inorganic Chemistry*. It is very concise and you will need to study it carefully to make sure you can remember all of the material.

How can I remember all this material?

Reading the booklet is an essential first step – but reading by itself is not a good way to get stuff into your memory. If you have bought the booklet and can write on it, you could try the following techniques to help you to memorise the material:

- underline or highlight the most important words in every paragraph
- underline or highlight scientific jargon – write a note of the meaning in the margin if you are unsure
- remember the number of items in a list – then you can tell if you have forgotten one when you try to remember it later
- tick sections when you are sure you know them – and then concentrate on the sections you do not yet know.

How can I check my progress?

The module test at the end is a useful check on your progress – you may want to wait until you have nearly completed the module and use it as a mock exam or try questions one by one as you progress. The answers show you how much you need to do to get the marks.

What if I get stuck?

The colour textbook *Collins Advanced Modular Sciences: Chemistry A2* is designed to support your A2 course. It provides more explanation than this booklet. It may help you to make progress if you get stuck.

Any other good advice?

- You will not learn well if you are tired or stressed. Set aside time for work (and play!) and try to stick to it.
- Don't leave everything until the last minute – whatever your friends may tell you it doesn't work.
- You are most effective if you work hard for shorter periods of time and then take a (short!) break. 30 minutes of work followed by a five or ten minute break is a useful pattern. Then get back to work.
- Some people work better in the morning, some in the evening. Find out which works better for you and do that whenever possible.
- Do not suffer in silence – ask friends and your teacher for help.
- Stay calm, enjoy it and ... good luck!

There are rigorous definitions of the main terms used in your examination – memorise these exactly.

The examiner's notes are always useful – make sure you read them because they will help with your module test.

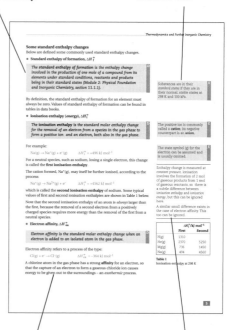

The main text gives a very concise explanation of the ideas in your course. You must study all of it – none is spare or not needed.

Further explanation references give a little extra detail, or directs you to other texts if you need more help or need to read around a topic.

14.1 Thermodynamics

14.1.1 *Enthalpy change, ΔH*

Definition

> **D** Enthalpy change, ΔH, is the amount of heat taken in or given out during any change (physical or chemical) under conditions of constant pressure.

See also *Module 2: Foundation Physical and Inorganic Chemistry,* section 11.1.1.

Standard enthalpy change, $\Delta H^{\ominus}$

The size of any enthalpy change depends on the amount of substance used as well as the conditions of measurement. Chemists are agreed in using **standard amounts** and **standard conditions** in order to be able to make useful comparisons between different measurements. The symbols for standard quantities are always followed by the superscript 'plimsoll' sign, $\ominus$.

Amount of substance

The standard amount used by chemists is the **mole** (*Module 1: Atomic Structure Bonding and Periodicity*, section 10.2.2). That this standard amount is being used will be obvious from the stated units, which will almost invariably be quoted as 'per mole' using the symbol **mol^{-1}**.

Standard conditions

The size of an enthalpy change for a given reaction also alters if either the pressure or the temperature at which it is measured is altered, even if reference is made to the standard amount (1 mol). In order to make comparison between different sets of data consistent with each other, it is convenient to quote enthalpy changes measured under agreed **standard conditions**.

> **D** **Standard pressure** The standard pressure chosen is 100 kPa (1 bar).
>
> **Standard temperature** The most common reference temperature used is 298 K.

Once the temperature is specified, the enthalpy change becomes the **standard enthalpy change**, $\Delta H^{\ominus}$(298 K) (*Module 2: Foundation Physical and Inorganic Chemistry,* section 11.1.1). If no temperature is indicated, then it is assumed that the reference temperature is 298 K. Thus, $\Delta H^{\ominus}$ on its own is the same as $\Delta H^{\ominus}$(298 K).

E **Molar quantities.** Enthalpy changes vary with **the amount** of substance present. Chemists use molar quantities and thus refer to enthalpy change per mole of *chemically balanced equation* for the process involved.

Some standard enthalpy changes

Below are defined some commonly used standard enthalpy changes.

- **Standard enthalpy of formation, $\Delta H_f^{\ominus}$**

> **D**
>
> The **standard enthalpy of formation** is the enthalpy change involved in the production of one mole of a compound from its elements under standard conditions, reactants and products being in their standard states (Module 2: Physical Foundation and Inorganic Chemistry, section 11.1.1).

> **E**
>
> Substances are in their standard states if they are in their normal, stable states at 298 K and 100 kPa.

By definition, the standard enthalpy of formation for an element must always be zero. Values of standard enthalpy of formation can be found in tables in data books.

- **Ionisation enthalpy (*energy*), $\Delta H_i^{\ominus}$**

> **D**
>
> The **ionisation enthalpy** is the standard molar enthalpy change for the removal of an electron from a species in the gas phase to form a positive ion and an electron, both also in the gas phase.

> **E**
>
> The positive ion is commonly called a **cation**; its negative counterpart is an **anion**.

For example:

$$Na(g) \rightarrow Na^+(g) + e^-(g) \qquad \Delta H_i^{\ominus} = +494 \text{ kJ mol}^{-1}$$

> **E**
>
> The state symbol (g) for the electron can be assumed and is usually omitted.

For a neutral species, such as sodium, losing a single electron, this change is called the **first ionisation enthalpy**.

The cation formed, $Na^+(g)$, may itself be further ionised, according to the process

$$Na^+(g) \rightarrow Na^{2+}(g) + e^- \qquad \Delta H_i^{\ominus} = +4560 \text{ kJ mol}^{-1}$$

which is called the **second ionisation enthalpy** of sodium. Some typical values of first and second ionisation enthalpies are shown in Table 1 below.

Note that the second ionisation enthalpy of an atom is *always* larger than the first, because the removal of a second electron from a positively charged species requires more energy than the removal of the first from a neutral species.

Enthalpy change is measured at *constant pressure*. Ionisation involves the formation of 2 mol of gaseous products from 1 mol of gaseous reactants, so there is a subtle difference between *ionisation enthalpy* and *ionisation energy*, but this can be ignored here.

A similar small difference exists in the case of electron affinity. This too can be ignored.

- **Electron affinity, $\Delta H_{ea}^{\ominus}$**

> **D**
>
> **Electron affinity** is the standard molar enthalpy change when an electron is added to an isolated atom in the gas phase.

Electron affinity refers to a process of the type:

$$Cl(g) + e^- \rightarrow Cl^-(g) \qquad \Delta H_{ea}^{\ominus} = -364 \text{ kJ mol}^{-1}$$

A chlorine atom in the gas phase has a strong **affinity** for an electron, so that the capture of an electron to form a gaseous chloride ion causes energy to be given out to the surroundings – an *exothermic* process.

	$\Delta H_i^{\ominus}/\text{kJ mol}^{-1}$	
	First	**Second**
H(g)	1310	
He(g)	2370	5250
Mg(g)	736	1450
Na(g)	494	4560

Table 1
Ionisation enthalpy at 298 K

	$\Delta H_{ea}^{\ominus}/kJ\ mol^{-1}$
H(g)	−72
F(g)	−348
Cl(g)	−364
Br(g)	−342
O(g)	−142

Table 2
Electron affinity at 298 K

E   A free radical is a species which results from the **homolytic fission** of a covalent bond. It contains an unpaired electron, since **homolytic fission** results in the splitting of the electron pair in a covalent bond, one electron to each partner.

$\Delta H_{diss}^{\ominus}/kJ\ mol^{-1}$			
H–H	436	Br–Br	194
H–Br	366	Cl–Cl	242
H–Cl	431	F–F	155
H–F	565	N≡N	945
H–OH	492	O=O	497

Table 3
Bond dissociation enthalpy

	$\Delta H_{at}^{\ominus}/kJ\ mol^{-1}$
C(graphite)	715
Na(s)	109
K(s)	90
Mg(s)	150

Table 4
Enthalpy of atomisation

Some typical values of electron affinity are shown in Table 2.

The enthalpy of formation of $O^{2-}(g)$ from $O^-(g)$ is +844 kJ mol^{-1}. The process is strongly endothermic since it takes a lot of energy to force a second electron onto the already negative O^- ion.

● **Bond dissociation enthalpy, $\Delta H_{diss}^{\ominus}$**

D   The **bond dissociation enthalpy** is the standard molar enthalpy change which accompanies the breaking of a covalent bond in a gaseous molecule to form two free radicals, also in the gas phase.

In order to show that free radicals are species each of which has one unpaired electron, it is considered correct, particularly in processes involving bond fission in a mass spectrometer or in radical chain reactions, to write a *dot* alongside the odd-electron species to indicate the unpaired electron. This species is called a **free radical**, e.g. the methyl radical $CH_3^{\bullet}$ or the chlorine atom $Cl^{\bullet}$.

In thermodynamic equations involving bond fission, it is quite common to omit the dots representing the unpaired electrons, since their presence is obvious from the equation as written, e.g.

$$Cl_2(g) \rightarrow 2Cl(g) \qquad \Delta H_{diss}^{\ominus} = +242\ kJ\ mol^{-1}$$

or $\quad CH_4(g) \rightarrow CH_3(g) + H(g) \qquad \Delta H_{diss}^{\ominus} = +435\ kJ\ mol^{-1}$

Some bond dissociation enthalpies for homolytic fission in a selection of covalent compounds are shown in Table 3.

● **Enthalpy of atomisation, $\Delta H_{at}^{\ominus}$**

D   The **enthalpy of atomisation** is the standard enthalpy change which accompanies the formation of one mole of gaseous atoms.

For an atomic solid, such as an element, the standard enthalpy of atomisation is simply the **standard enthalpy of sublimation** of the solid. For example:

$$Na(s) \rightarrow Na(g) \qquad \Delta H_{sub}^{\ominus} = +107\ kJ\ mol^{-1}$$

In such a case, the enthalpy of atomisation is the same as the enthalpy of sublimation:

$$\Delta H_{at}^{\ominus} = \Delta H_{sub}^{\ominus}$$

Enthalpies of atomisation (sublimation) for a selection of substances are shown in Table 4. Sublimation always requires an input of energy (*endothermic* process), so these enthalpies are all positive.

In the case of bond fission, a diatomic molecule will produce two moles of atoms, so the enthalpy of atomisation is half the bond dissociation enthalpy. Thus, for chlorine:

$$\tfrac{1}{2}Cl_2(g) \rightarrow Cl(g) \qquad \Delta H_{at}^{\ominus} = +121\ kJ\ mol^{-1}$$

so $\quad \Delta H_{at}^{\ominus} = \tfrac{1}{2}\Delta H_{diss}^{\ominus}$

- **Lattice enthalpy, $\Delta H_L^\ominus$**

> **D**
>
> The **enthalpy of lattice dissociation** is the standard enthalpy change which accompanies the separation of one mole of a solid ionic lattice into its gaseous ions.

For example:

$$NaCl(s) \rightarrow Na^+(g) + Cl^-(g) \qquad \Delta H_L^\ominus = +771 \text{ kJ mol}^{-1}$$

If **lattice dissociation** is used as a defining equation, as above, all lattice enthalpies are *positive*, since all ionic crystals are energetically more favoured than their separated gaseous ions. Consequently, it requires an *input* of energy to disrupt the crystal lattice and form separated gaseous ions.

If the reverse defining equation is used

$$Na^+(g) + Cl^-(g) \rightarrow NaCl(s) \qquad \Delta H_L^\ominus = -771 \text{ kJ mol}^{-1}$$

then the process of interest is **lattice formation**, and the resulting **enthalpy of lattice formation** is always *negative*. Be aware that such conflicting definitions exist and be able to distinguish which definition is in use from the sign of the resulting lattice enthalpy or from the direction of the arrow in the defining equation.

> **E**
>
> Bond *making* is exothermic.
> Bond *breaking* is endothermic.

Some typical lattice enthalpy values are shown in Table 5.

- **Enthalpy of hydration, $\Delta H_{hyd}^\ominus$**

> **D**
>
> The **enthalpy of hydration** is the standard molar enthalpy change for the process
>
> $$X^\pm(g) \xrightarrow{\text{water}} X^\pm(aq) \qquad \Delta H^\ominus = \Delta H_{hyd}^\ominus$$

$\Delta H_L^\ominus$/kJ mol^{-1}	
NaF(s)	902
NaCl(s)	771
KCl(s)	701
MgO(s)	3889
MgS(s)	3238

Table 5
Lattice enthalpy at 298 K

The single symbol, $X^\pm(g)$ is used to indicate either a gaseous cation, $X^+(g)$, or a gaseous anion, $X^-(g)$, depending on the appropriate hydrated species in solution.

Hydration enthalpies for a selection of ions are shown in Table 6.

- **Enthalpy of solution, $\Delta H_{sol}^\ominus$**

> **D**
>
> The **enthalpy of solution** is the standard enthalpy change for the process in which one mole of an ionic solid dissolves in an amount of water large enough to ensure that the dissolved ions are well separated and do not interact with one another.

$\Delta H_{hyd}^\ominus$/kJ mol^{-1}			
Li$^+$	−519	F$^-$	−506
Na$^+$	−406	Cl$^-$	−364
K$^+$	−322	Br$^-$	−335

Table 6
Hydration enthalpy for individual ions

For example:

$$NaCl(s) \xrightarrow{\text{water}} Na^+(aq) + Cl^-(aq) \qquad \Delta H_{sol}^\ominus = +2 \text{ kJ mol}^{-1}$$

The Born–Haber cycle

The enthalpy of formation of a solid ionic compound can be broken up into a number of steps, arranged in a cycle called the **Born–Haber cycle**. This cycle includes a step involving the enthalpy of lattice formation, $\Delta H_L^\ominus$, so that it is possible to deduce the value of the lattice enthalpy from other thermodynamic data using the Born–Haber cycle. The procedure used in the case of sodium chloride is shown in Fig 1.

Fig 1

A Born–Haber cycle for the determination of the lattice enthalpy of sodium chloride. The sum of the enthalpy changes round the cycle is zero, so that the magnitude of any unknown enthalpy in the cycle can be found if all the other enthalpies are known.

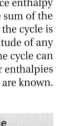

E

The Born–Haber cycle

Starting from the elements, proceed to form the compound by just 1 step (directly) or by several (indirectly).

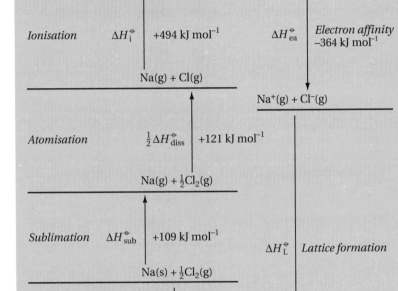

Applying Hess's law:

$$\Delta H^\ominus \text{(single step)} = \Sigma\Delta H^\ominus \text{ (five steps)}$$

so
$$\Delta H_f^\ominus = \Delta H_{sub}^\ominus + \tfrac{1}{2}\Delta H_{diss}^\ominus + \Delta H_i^\ominus + \Delta H_{ea}^\ominus + \Delta H_L^\ominus$$

and
$$\Delta H_L^\ominus = -\Delta H_{sub}^\ominus - \tfrac{1}{2}\Delta H_{diss}^\ominus - \Delta H_i^\ominus - \Delta H_{ea}^\ominus + \Delta H_f^\ominus$$

therefore $\Delta H_L^\ominus(\text{NaCl}) = -\ 109\ -\ 121\ -\ 494\ -(-364\ ) + (-411)$

$$= -771 \text{ kJ mol}^{-1}$$

The standard enthalpy of formation of NaCl(s) is -411 kJ mol^{-1}.

Example 1

Calculate the lattice enthalpy of sodium bromide. Use data taken from tables in the text. The enthalpy of vaporisation of bromine is +30 kJ mol⁻¹ and the standard enthalpy of formation of solid sodium bromide is −360 kJ mol⁻¹.

Method

Construct a cycle, as above. Because bromine is a liquid at 298 K, an extra equation/step is needed.

Answer

$\Delta H_L^{\ominus}(\text{NaBr}) = -109 - 15 - 97 - 494 + 342 - 360 = -733 \text{ kJ mol}^{-1}$

Calculating enthalpies of solution

Example 2 shows a calculation which uses enthalpies of hydration and lattice enthalpy.

Example 2

Calculate the enthalpy of solution of sodium chloride using data taken from tables in the text.

Answers

Method 1

Draw an enthalpy cycle as shown below.

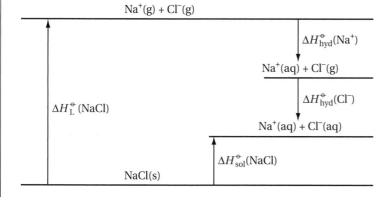

Hence

$$\Delta H_{sol}^{\ominus}(\text{NaCl}) = \Delta H_L^{\ominus}(\text{NaCl}) + \Delta H_{hyd}^{\ominus}(\text{Na}^+) + \Delta H_{hyd}^{\ominus}(\text{Cl}^-)$$

Therefore

$$\Delta H_{sol}^{\ominus}(\text{NaCl}) = \quad +771 \quad + \quad (-405) \quad + \quad (-364) \quad = +2 \text{ kJ mol}^{-1}$$

E The enthalpy of solution is very small, and is calculated here as the difference between two very large quantities. Such calculations are often very unreliable and are subject to round-off errors, particularly here where the other enthalpies are quoted to only three significant figures. Note that their difference is good to only one significant figure.

Example 2 (continued)

Method 2

Write equations, including enthalpy changes (reversed where appropriate) for lattice enthalpy and hydration enthalpies in such a way that simple addition yields the equation for enthalpy of solution. Addition of the enthalpy changes gives the overall enthalpy of reaction.

$$NaCl(s) \xrightarrow{\text{water}} Na^+(g) + Cl^-(g) \qquad \Delta H^{\ominus}_L = +771 \text{ kJ mol}^{-1}$$

$$Na^+(g) \xrightarrow{\text{water}} Na^+(aq) \qquad \Delta H^{\ominus}_{hyd} = -405 \text{ kJ mol}^{-1}$$

$$Cl^-(g) \xrightarrow{\text{water}} Cl^-(aq) \qquad \Delta H^{\ominus}_{hyd} = -364 \text{ kJ mol}^{-1}$$

$$NaCl(s) \xrightarrow{\text{water}} Na^+(aq) + Cl^-(aq) \qquad \Delta H^{\ominus}_{sol} = +2 \text{ kJ mol}^{-1}$$

Comment

Either method may be used with confidence. The one that is chosen is entirely a matter of personal preference.

Calculating enthalpy changes using mean bond enthalpies

When a reaction occurs, bonds in molecules are broken (requiring an input of heat energy from the surroundings) and new bonds are formed (causing a flow of heat energy into the surroundings). The overall enthalpy change for the reaction represents the net balance between heat energy taken in and heat energy evolved, between old bonds broken and new ones formed.

The strengths of bonds can be determined using spectroscopy, by measuring the **bond dissociation enthalpy**, $\Delta H^{\ominus}_{diss}$ (X–Y). This quantity is the enthalpy change accompanying the fission of a bond in a gas-phase species, X–Y, to form two new gas-phase fragments:

$$X-Y(g) \rightarrow X(g) + Y(g) \qquad \Delta H^{\ominus} = \Delta H^{\ominus}_{diss} (X-Y)$$

E The dissociation enthalpy of a given bond (O–H, say) depends on the structure of the rest of the molecule around it. To break an O–H bond in ethanol requires a different input of heat energy than that needed to break such a bond in ethanoic acid.

If all **bond enthalpies** were known, it would be possible to predict overall enthalpy changes simply by summing over those bonds that had altered, writing a positive enthalpy contribution for any bond broken and a negative one for any bond formed.

Unfortunately, such a simplistic approach fails. It is clear why this is so if the case of water is considered. Water has two entirely equivalent O–H bonds. Yet, breaking the first:

$$H_2O(g) \rightarrow H(g) + OH(g) \qquad \Delta H^{\ominus}_{diss} (HO-H) = +492 \text{ kJ mol}^{-1}$$

gives rise to a bigger enthalpy change than breaking the second.

$$HO(g) \rightarrow H(g) + O(g) \qquad \Delta H^{\ominus}_{diss} (O-H) = +428 \text{ kJ mol}^{-1}$$

Changes in the environment of a bond change its strength. In methanol, for example:

$$CH_3OH(g) \rightarrow CH_3O(g) + H(g) \qquad \Delta H^{\ominus}_{diss} (CH_3O-H) = +437 \text{ kJ mol}^{-1}$$

the bond enthalpy corresponds to the breaking of neither of the OH bonds in water.

The solution to this difficulty lies in compiling a list of **mean bond enthalpies**, $B(X-Y)$, which were considered previously in *Module 2: Foundation Physical and Inorganic Chemistry*, section 11.1.4. Some values are shown in Table 7. By averaging bond dissociation enthalpies measured in a wide variety of different compounds, mean values can be found.

Mean bond enthalpies provide a simple way of calculating approximate values of overall enthalpy changes, particularly where other data are not available. However, the values obtained are only *approximate* and should always be used with caution. An illustration of the use of mean bond enthalpies is given in Example 3.

The **mean bond enthalpy** of a bond X–Y is given the symbol $B(X-Y)$; it is an average value and can be used in *approximate* calculations.

B(X–Y)/kJ mol^{-1}			
H–H	436	C–Cl	328
H–C	413	C–N	305
H–N	391	C–O	360
H–O	463	C=O	743
C–C	348	N–O	158
C=C	612	O–O	146
C≡C	838	O=O	498

Table 7
Mean bond enthalpies

Example 3

Calculate the standard enthalpy change for the complete combustion of propane. Use the mean bond enthalpy data given in Table 7.

Method

Firstly, write an equation for the overall reaction. Next, *'atomise'* the reactant molecules by breaking all the bonds present on the reactant side of the equation.

Then *'reassemble'* the product molecules from the gaseous atoms by making all the bonds present on the product side of the equation. The overall enthalpy change is the sum of the enthalpy needed to break the bonds and the enthalpy released when bonds are made.

Enthalpy change = energy required for breaking bonds
+ energy released from making bonds

Answer
The overall equation is:

$$C_3H_8(g) + 5O_2(g) \longrightarrow 3CO_2(g) + 4H_2O(g)$$

Draw an enthalpy cycle as shown below.

Bond *breaking* is an *endothermic* process (ΔH is positive).

Bond *making* is an *exothermic* process (ΔH is negative).

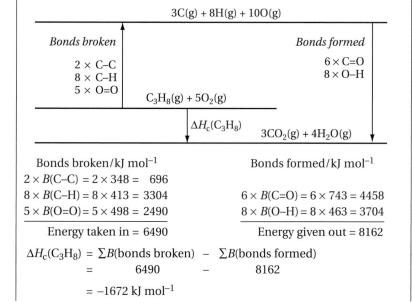

$$3C(g) + 8H(g) + 10O(g)$$

Bonds broken

$2 \times$ C–C
$8 \times$ C–H
$5 \times$ O=O

$C_3H_8(g) + 5O_2(g)$

Bonds formed

$6 \times$ C=O
$8 \times$ O–H

$\Delta H_c(C_3H_8)$ $3CO_2(g) + 4H_2O(g)$

Bonds broken/kJ mol^{-1}
$2 \times B(\text{C–C}) = 2 \times 348 = 696$
$8 \times B(\text{C–H}) = 8 \times 413 = 3304$
$5 \times B(\text{O=O}) = 5 \times 498 = 2490$

Energy taken in = 6490

Bonds formed/kJ mol^{-1}
$6 \times B(\text{C=O}) = 6 \times 743 = 4458$
$8 \times B(\text{O–H}) = 8 \times 463 = 3704$

Energy given out = 8162

$$\begin{aligned}\Delta H_c(C_3H_8) &= \Sigma B(\text{bonds broken}) - \Sigma B(\text{bonds formed})\\ &= \qquad 6490 \qquad - \qquad 8162 \\ &= -1672 \text{ kJ mol}^{-1}\end{aligned}$$

E The value found using mean bond enthalpies is within 10% of the correct (*calorimetric*) value, which is typical of the level of accuracy that can be expected from this method. In cases where calorimetric data are not available, *mean bond enthalpy* values can provide a satisfactory estimate of inaccessible reaction enthalpy data.

Example 3 (continued)

Comment

The value calculated using mean bond enthalpies is only *approximate* because the **mean bond enthalpy** data used refer to *average* values of bond enthalpies derived from measurements in many different compounds.

14.1.2 *Free-energy change, ΔG, and entropy change, ΔS*

Spontaneous change

A **spontaneous** change is one which has a natural tendency to occur without being driven by any external influences. Spontaneous changes are familiar from everyday life. The air compressed in a bicycle tyre escapes spontaneously if the valve is removed, and a lot of physical effort is needed to pump it up again. Hot soup cools spontaneously, and heat energy from a gas ring is needed to warm it up again. Ice cream melts spontaneously on a hot day, and needs the electrical energy put into a refrigerator to freeze it again. Iron rusts spontaneously in damp air, and it takes all the vast energy of a blast-furnace to get iron back from an oxide ore. A rechargeable dry cell can light the bulb of a torch spontaneously, but has to be left to draw energy from the mains when being recharged.

E If a process in one direction is spontaneous, the reverse will require an input of external energy.

Left to themselves, therefore, some things simply happen. Others don't happen unless energy is expended. A **spontaneous change** is one which can occur in one particular direction but not in reverse (unless conditions such as temperature are changed).

The enthalpy factor

Many spontaneous chemical reactions appear to be driven by a favourable change in enthalpy. In all the changes listed above, it would be easy to reach the simple (but mistaken) conclusion that spontaneous change occurs if heat is given out as the change is taking place. But this is not the whole story.

Exothermic reactions are often spontaneous. The natural direction of spontaneous change is from higher to lower enthalpy, with an (exothermic) release of the difference in energy between the two. For example, the reaction:

$$H_2(g) + \tfrac{1}{2}O_2(g) \rightarrow H_2O(g) \qquad \Delta H^{\ominus} = -242 \text{ kJ mol}^{-1}$$

E Most spontaneous reactions are exothermic. But some endothermic reactions are also spontaneous.

liberates vast quantities of heat energy. The reaction is exothermic, the enthalpy change is negative ($\Delta H < 0$), and this favours spontaneous reaction.

It would be tempting to conclude that spontaneous reactions must involve a release of heat energy. However, some reactions are spontaneous even though they are **endothermic**. For example, the reaction:

$$KHCO_3(s) + HCl(aq) \rightarrow KCl(aq) + H_2O(l) + CO_2(g) \quad \Delta H^\ominus = +25 \text{ kJ mol}^{-1}$$

is endothermic and the temperature of the reaction mixture drops when the reactants are mixed. Yet it proceeds spontaneously. Clearly, there must be some additional factor that causes reactions to occur, over and above the simple release of heat energy.

The entropy factor

The additional factor that helps to drive spontaneous change in a given direction is called the **entropy**, which is given the symbol **S**.

Before proceeding in the next sections to relate entropy to everyday experience, it is useful to make a few statements about entropy and its nature, just to get an insight into this new quantity that turns out to be one of the driving forces in spontaneous change.

(a) Entropy and disorder

It proves very helpful to *think* of entropy in terms of *disorder*. An increase in *entropy* can be visualised as an increase in *disorder*. Processes leading to increased chaos are rather more likely than those leading to order. Heating leads to an increase in disorder among molecules, which fits with the increase in entropy as a substance changes from solid, to liquid, to gas.

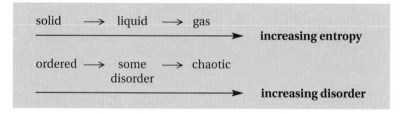

(b) Variation of entropy with temperature

A graph showing typical changes of entropy as temperature is increased is shown in Fig 2.

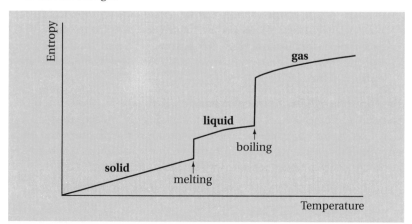

Fig 2
Variation of entropy with temperature for a typical substance.

Entropy, S, has the units J K^{-1} mol^{-1}.

It is clear that:

- entropy increases slowly with temperature in solids, liquids and gases

- a **phase change** (*change of physical state*) causes a sudden change in entropy

- boiling causes a larger increase in entropy than does melting

(c) Magnitude of entropy in different substances

In Table 8, some standard entropies of different substances are listed. Two broad trends can be noted:

- simple molecules generally seem to have lower entropies than more complicated molecules, e.g.

$$S^{\ominus}(CO) < S^{\ominus}(CO_2); \quad S^{\ominus}(He) < S^{\ominus}(O_2); \quad S^{\ominus}(NaCl) < S^{\ominus}(NaHCO_3)$$

- for substances of similar complexity, entropy generally increases on going from solid, to liquid, to gas, e.g.

$$S^{\ominus}(H_2O(s)) < S^{\ominus}(H_2O(l)) < S^{\ominus}(H_2O(g))$$

$$S^{\ominus}(CH_3OH(l)) < S^{\ominus}(CH_3OH(g))$$

$S^{\ominus}_{298}$/kJ mol^{-1}			
diamond	2.4	He(g)	126
graphite	5.7	Ar(g)	155
NaCl(s)	72	O_2(g)	205
NaHCO$_3$(s)	102	CO(g)	198
SiO$_2$(s)	42	CO$_2$(g)	214
H$_2$O(l)	70	H$_2$O(g)	189
CH$_3$OH(l)	127	CH$_3$OH(g)	240

Table 8
Standard entropy at 298 K

(d) Units in which entropy is measured

In the SI system, the units are J K^{-1} mol^{-1}. In themselves, these units convey little more about entropy, but will prove useful and revealing later on when the separate influences of enthalpy and entropy are combined in determining the direction of spontaneous change.

(e) Absolute entropies, $S^{\ominus}$

Since entropy is so closely linked to disorder, and decreases as the temperature is lowered, it is reasonable to suggest that at 0K, absolute zero temperature, all disorder will have vanished and all substances will be perfectly ordered and have *zero entropy*. This turns out to be true for most substances, especially perfectly-ordered crystals.

Absolute entropy values are based on the premise that all substances are perfectly ordered and have zero entropy at the absolute zero of temperature.

The concept of zero entropy has one important consequence. Unlike enthalpy, for which it is only possible to consider standard enthalpy *changes*, $\Delta H^{\ominus}$ (i.e. *differences between enthalpy values*), in the case of entropy there is a well-defined starting point to the entropy scale (zero entropy at zero kelvins). As a consequence, it is possible to speak of and to list values for **absolute standard entropies**, $S^{\ominus}$, and from these to calculate **standard entropy changes**, $\Delta S^{\ominus}$. Table 8 lists some **absolute entropy values**.

Entropy change in chemical reactions, $\Delta S^{\ominus}$

Standard entropy changes in systems undergoing chemical change are calculated as the difference between the entropy of the products and the entropy of the reactants. Absolute entropy values of products and reactants are obtained from tables of **standard entropies** such as Table 8.

Entropy calculations using standard entropy values use the equation:

$$\Delta S^{\ominus} = \Sigma S^{\ominus}_{products} - \Sigma S^{\ominus}_{reactants}$$

Two illustrations (Examples 4 and 5) show how such calculations are performed.

Example 4

Calculate the entropy change which accompanies the combustion of graphite. Use the entropy data given in Table 8.

Method

The reaction in question is:

$$C(s) + O_2(g) \rightarrow CO_2(g)$$

Answer

$$\Delta S^\ominus = \Sigma S^\ominus_{products} - \Sigma S^\ominus_{reactants}$$

$\Sigma S^\ominus_{products}$ $= 214 \text{ J K}^{-1} \text{ mol}^{-1}$

$\Sigma S^\ominus_{reactants}$ $(5.7 + 205) = 210.7 \text{ J K}^{-1} \text{ mol}^{-1}$

Therefore $\Delta S^\ominus = +3.3 \text{ J K}^{-1} \text{ mol}^{-1}$

Comment

The entropy change here is quite small. Carbon dioxide and oxygen are both gases and have similar entropies. Graphite is a very low-entropy solid. So even though two moles form one, the number of gas moles does not change and there is not much change in entropy.

Example 5

Calculate the entropy change that accompanies the decomposition by heating of solid sodium hydrogencarbonate. Use the entropy data given in Table 8 and the additional information that solid sodium carbonate has a standard entropy of 136 J K^{-1} mol^{-1}.

Method

The reaction in question is:

$$2NaHCO_3(s) \rightarrow Na_2CO_3(s) + H_2O(g) + CO_2(g)$$

Answer

$$\Delta S^\ominus = \Sigma S^\ominus_{products} - \Sigma S^\ominus_{reactants}$$

$\Sigma S^\ominus_{products}$ $136 + 189 + 214 = 539 \text{ J K}^{-1} \text{ mol}^{-1}$

$\Sigma S^\ominus_{reactants}$ $2 \times 102 = 204 \text{ J K}^{-1} \text{ mol}^{-1}$

Therefore $\Delta S^\ominus = +335 \text{ J K}^{-1} \text{ mol}^{-1}$

Comment 1

The entropy change calculated is *per mole of equation as written*, which here refers to the decomposition of 2 mol of NaHCO$_3$. For one mole of NaHCO$_3$, ΔS is halved, giving 167.5 J K^{-1} mol^{-1}.

Comment 2

The entropy change here is very much larger than that found in Example 4. In this case, 2 mol of gas are formed from a solid, so the disorder of the system is increased much more.

Gibbs free-energy change, $\Delta G^{\ominus}$, spontaneous change and feasibility

It is clear that reactions in which there is a release of heat energy (*exothermic processes*) tend to happen quite often. Equally, reactions in which there is an intake of heat energy (*endothermic processes*) but which lead to an increase in disorder (*increase in entropy*) also happen quite often. The sometimes conflicting demands of enthalpy and entropy are brought together in the relationship

$$\Delta G^{\ominus} = \Delta H^{\ominus} - T\Delta S^{\ominus}$$

where the free-energy change, ΔG, combines the influence of both enthalpy and entropy.

This equation expresses the benefit of an exothermic reaction (ΔH *negative*) with one in which entropy increases (ΔS *positive*), leading to the conclusion that:

> **D** $\Delta G^{\ominus}$ must be negative (or zero) for feasible change.

Spontaneous ≡ feasible

In thermodynamics, the words '*spontaneous*' and '*feasible*' have exactly the same meaning. In everyday speech, however, the word spontaneous (as in 'spontaneous applause' or 'spontaneous tears') implies something that is almost *inevitable*, that simply *must* happen. In thermodynamics, spontaneity has to do only with a *tendency* for something to happen, and not if it *actually will* happen. Although the term '*feasible*' is to be preferred, implying as it does, something that is possible but not in inevitable, the term '*spontaneous*' is firmly enshrined in the language of equilibrium and provides an accepted alternative to *feasible*.

Quite often, the speed at which a feasible change occurs can be rather slow (as in the cooling of a cup of coffee) or even infinitely slow (as in the change from diamond into graphite). Nonetheless, both these changes (cold coffee and no diamonds) are truly *feasible* because a decrease in free energy occurs.

Thermodynamics has nothing to say about the speed with which things happen, only if they *can* happen. Thermodynamics answers the first and fundamental question: *can it go?* (is the reaction *feasible*, is $\Delta G < 0$?) Only if the answer to this first question is '*Yes*' is it worth posing the second question: how fast will it go? (are the *kinetics* favourable?).

A chemical or physical change is said to be *feasible* if the value of ΔG is *negative* or *zero*. When ΔG is *positive*, the change is said to be *unfeasible* (or *not feasible*).

E The reaction between gaseous hydrogen and oxygen at room temperature is *spontaneous/feasible* because ΔG for the formation of water from its elements is very negative. So it *can* go. The reaction will not go, however, unless a spark or a flame is applied, whereupon an explosion results.

Similarly, hydrogen and chlorine do not react in the dark at room temperature but do so with explosive violence in sunlight.

In such cases, the reaction needs to overcome an **activation energy** barrier (see *Module 2: Foundation Physical and Inorganic Chemistry*, section 11.2.3) before reaction can occur.

E Reactions for which ΔG is only slightly positive (i.e. just on the wrong side of zero and therefore unfeasible, but not greatly so) can reach an equilibrium which favours the reactants but still has an appreciable concentration of products present. Manufacturers are frequently forced to take advantage of such unfeasible reactions.

Calculations involving $\Delta G^\ominus$

From standard enthalpy and standard entropy data, the standard free-energy change can be calculated using $\Delta G^\ominus = \Delta H^\ominus - T\Delta S^\ominus$ and hence the feasibility of chemical reactions can be determined. This is illustrated in Examples 6 to 8 that follow. The data required are given in Table 9.

Chemical reaction	$\Delta H^\ominus_{298}/$ kJ mol^{-1}	$\Delta S^\ominus_{298}/$ J K^{-1} mol^{-1}
$C(s) + O_2(g) \rightarrow CO_2(g)$	−394	+3.3
$2Fe(s) + \frac{3}{2}O_2(g) \rightarrow Fe_2O_3(s)$	−825	−272
$2NaHCO_3(s) \rightarrow Na_2CO_3(s) + H_2O(g) + CO_2(g)$	+130	+335

E Note that the first reaction is *always* feasible, that the second is favoured by *low* temperatures, while the third is favoured by *high* temperatures.

Table 9
Standard enthalpy and entropy changes for three chemical reactions

---

Example 6

Calculate the standard free-energy change for the combustion of graphite at 298 K. Use data given in Table 9.

Reaction $C(s) + O_2(g) \rightarrow CO_2(g)$

Answer

$\Delta G^\ominus = \Delta H^\ominus - T\Delta S^\ominus$

$\Delta G^\ominus_{298} = -394 \text{ kJ mol}^{-1} - \dfrac{298 \times 3.3}{1000} \text{ kJ mol}^{-1}$

$= -395 \text{ kJ mol}^{-1}$

Comment

The energy units of ΔH and $T\Delta S$ must be made compatible, which renders the entropy term (3.3×10^{-3} kJ K^{-1} mol^{-1}) vanishingly small, even when multiplied by 298 K.

E ΔG is negative, so the reaction is *feasible*. The entropy term is very small so $\Delta G \approx \Delta H$. Because ΔS is so small, ΔG will not vary much with temperature, though in practice the reaction is extremely slow at 298 K (high activation energy).

Example 7

Calculate the standard free-energy change for the rusting of iron at 298 K. Use data given in Table 9.

Reaction $2Fe(s) + \frac{3}{2}O_2(g) \rightarrow Fe_2O_3(s)$

Answer

$\Delta G^\ominus = \Delta H^\ominus - T\Delta S^\ominus$

$\Delta G^\ominus_{298} = -825 \text{ kJ mol}^{-1} - \dfrac{298 \times (-272)}{1000} \text{ kJ mol}^{-1}$

$= -744 \text{ kJ mol}^{-1}$

Comment

ΔG is hugely negative, so the rusting of iron is highly feasible at room temperature, in complete accord with everyday experience. The reaction is favoured by low temperatures but does not cease to be feasible until the temperature exceeds 3000 K. It is strongly driven by a very exothermic enthalpy change.

E ΔG here is very negative, so the reaction is feasible. The entropy change is negative because disorder in the gas is lost forming a solid. ΔG is negative because the exothermic ΔH dominates.

E ΔG is positive, so at 298 K the reaction is *not feasible*. However, ΔG is not very large, so raising the temperature will allow the $-T\Delta S$ term to dominate. The reaction will become *feasible* at higher temperatures. The entropy change is positive since one gaseous mole forms for the loss of half a mole of solid.

Example 8

Calculate the standard free-energy change for the decomposition of one mole of sodium hydrogencarbonate at 298 K. Use data given in Table 9.

Reaction $\quad\quad 2NaHCO_3(s) \rightarrow Na_2CO_3(s) + H_2O(g) + CO_2(g)$

Answer

$$\Delta G^{\ominus} = \Delta H^{\ominus} - T\Delta S^{\ominus}$$

$$\Delta G^{\ominus}_{298} = 130 \text{ kJ mol}^{-1} - \frac{298 \times 335}{1000} \text{ kJ mol}^{-1} = +30 \text{ kJ mol}^{-1}$$

so, for 1 mol of NaHCO$_3$, $\Delta G^{\ominus}_{298} = \frac{1}{2} \times 30 = +15$ kJ mol^{-1} (2 s.f.)

Comment 1
The unit *mol^{-1}* means *per mole of the equation as written*, which in this case involves 2 mol of NaHCO$_3$.

Comment 2
ΔG is only slightly positive at 298 K, so the decomposition of sodium hydrogencarbonate is *almost* feasible at room temperature, but needs some heat input to become *actually* feasible. Again, this is in accord with everyday experience. The reaction is favoured by high temperatures and needs a temperature close to 400 K to become feasible (see Example 11).

Entropy in physical changes
In many physical changes, there is an increase or decrease in order (and hence in entropy). The most obvious of these are the melting of a solid or the boiling of a liquid, and these two cases are considered in Examples 9 and 10, respectively.

Melting (fusion)
When ice melts at constant temperature (0 °C) and constant pressure (1 bar), the mixture of ice and water which results has *no spontaneous tendency* either to solidify (all ice) or to liquefy (all water) unless external influences (removal or addition of heat) are involved. Left in a system which prevents a flow of heat (e.g. a thermos flask) an ice/water mixture will remain at 0 °C as long as both ice and water are present. If a little heat leaks in, a little of the ice absorbs this heat and melts. The new equilibrium has slightly different amounts of ice and water present, but the temperature remains at 0 °C.

E The influence on the **position of equilibrium** of the addition of heat to, or the removal of heat from, melting ice provides a graphic example of **Le Chatelier's principle** in action. The system responds to nullify the influence of the removal or addition of heat energy, and does so very successfully, maintaining the temperature at a constant 0 °C.

Melting ice is an example of a **system at equilibrium** which has no tendency to move spontaneously in one direction or the other.

Since it is the sign of ΔG which determines the direction of change, it can be concluded that:

D *For a system at equilibrium, $\Delta G = 0$*

Applying the criterion $\Delta G = 0$ to the equation $\Delta G^{\ominus} = \Delta H^{\ominus} - T\Delta S^{\ominus}$

leads to $\quad\quad\quad\quad\quad\quad \Delta H^{\ominus} = T\Delta S^{\ominus}$

whence, for melting (*fusion*), $\quad \Delta S^{\ominus}_{fus} = \dfrac{\Delta H^{\ominus}_{fus}}{T_{fus}}$

Boiling (vaporisation)

Arguments similar to the ones used for melting apply here.

When water boils at constant temperature (100 °C) and constant pressure (1 bar), the mixture of water and steam which results has *no spontaneous tendency* either to liquefy (all water) or to vaporise (all steam) unless external influences (addition or removal of heat) are involved. Left in a system which prevents a flow of heat, a water/steam mixture will remain at 100 °C as long as both water and steam are present. If a little heat leaks in, a little of the water absorbs this heat and becomes steam. The new equilibrium will have slightly different amounts of water and steam present, but the temperature will remain at 100 °C.

Boiling water is also a **system at equilibrium** which has no tendency to move spontaneously in one direction or the other.

Again, it is the sign of ΔG which determines the tendency of a system to change, so it can be concluded that:

for boiling (*vaporisation*), $\Delta S_{vap}^{\ominus} = \dfrac{\Delta H_{vap}^{\ominus}}{T_{vap}}$

Illustrations are given in Examples 9 and 10 below.

E

Another example of **Le Chatelier's principle** in action:

When both phases are present in equilibrium at constant pressure, the temperature remains constant.

This property of equilibria is used in establishing two fixed points (0 °C and 100 °C) on the Celsius temperature scale.

Example 9

Calculate the entropy change which accompanies the melting of ice. The enthalpy of fusion of ice is 6.0 kJ mol^{-1}.

Method
The reaction in question is $H_2O(s) \rightleftharpoons H_2O(l)$
This is an equilibrium change of state at fixed temperature, so $\Delta G = 0$.

$$\Delta S_{fus}^{\ominus} = \frac{\Delta H_{fus}^{\ominus}}{T_{fus}}$$

Answer

$\Delta H_{fus}^{\ominus} = +6.0 \text{ kJ mol}^{-1}$ $T_{fus} = 273 \text{ K}$

Therefore $\Delta S_{fus}^{\ominus} = \dfrac{+6.0 \times 10^3 \text{ J mol}^{-1}}{273 \text{ K}} = +22 \text{ J K}^{-1} \text{ mol}^{-1}$ (*2 s.f.*)

Comment
ΔS is positive, as is to be expected when a solid melts to form a liquid. Note the need to convert enthalpy (in kJ mol^{-1}) into J mol^{-1} in order to give entropy in J K^{-1} mol^{-1}.

Example 10

Calculate the entropy change which accompanies the boiling of water. The enthalpy of vaporisation of water is 44.0 kJ mol^{-1}.

Method
The reaction in question is $H_2O(l) \rightleftharpoons H_2O(g)$

Example 10 (continued)

This is an equilibrium change of state at fixed temperature, so $\Delta G = 0$.

$$\Delta S^{\ominus}_{\text{vap}} = \frac{\Delta H^{\ominus}_{\text{vap}}}{T_{\text{vap}}}$$

Answer

$$\Delta H^{\ominus}_{\text{vap}} = +44.0 \text{ kJ mol}^{-1} \qquad T_{\text{vap}} = 373 \text{ K}$$

Therefore $\quad \Delta S^{\ominus}_{\text{vap}} = \dfrac{+44.0 \times 10^3 \text{ J mol}^{-1}}{373 \text{ K}} = +118 \text{ J K}^{-1} \text{ mol}^{-1}$

Comment

The entropy change here has a much more positive value than that for a solid melting to form a liquid. The increase in entropy is associated with the creation of one mole of disordered vapour molecules from one mole of relatively ordered liquid molecules.

Calculation of the temperature at which a reaction becomes feasible

A reaction which is not feasible at one temperature may become feasible if the temperature is raised (or lowered). The temperature at which feasibility is just achieved is that at which there is no spontaneous tendency for the reaction to go one way or the other. At this temperature, ΔG becomes zero and

$$\Delta H^{\ominus} = T\Delta S^{\ominus}$$

whence, for feasibility, $\qquad T = \dfrac{\Delta H^{\ominus}}{\Delta S^{\ominus}}$

The temperature of feasibility can then be calculated if ΔH and ΔS are both known. An illustration of such a calculation is given in Example 11 below.

Example 11

Calculate the temperature at which the thermal decomposition of sodium hydrogencarbonate becomes feasible. The standard enthalpy and entropy changes for the decomposition are given in Table 9.

Reaction $\qquad 2\text{NaHCO}_3(s) \rightleftharpoons \text{Na}_2\text{CO}_3(s) + \text{H}_2\text{O}(g) + \text{CO}_2(g)$

Answer

The reaction becomes feasible when $\Delta G = 0$

Therefore $\qquad \Delta H^{\ominus} = T\Delta S^{\ominus}$ and $T = \dfrac{\Delta H^{\ominus}}{\Delta S^{\ominus}}$

so $T = \dfrac{+130 \times 10^3}{+335} = 388 \text{ K}$

Comment

The decomposition reaction is feasible ($\Delta G \leq 0$) above 388 K. Below this temperature, the reverse reaction (the formation of sodium hydrogencarbonate) is feasible. The stoichiometry chosen for the decomposition does not affect the feasibility temperature. Halving the chemical equation (to involve only one mole of NaHCO_3) also halves both ΔH and ΔS, leading to the same feasibility temperature.

E Note that the energy parts of the units of ΔH (*kilojoules* per mole) and ΔS (*joules* per kelvin per mole) must be made compatible: this is done by converting kJ mol^{-1} into J mol^{-1} using the factor of 10^3.

E The reaction becomes more feasible as the temperature is raised. This outcome is in accord with the intuitive belief that sodium hydrogencarbonate is more likely to decompose on heating than on cooling!

In the example above, ΔH and ΔS are *both positive*, so the reaction is not feasible at low temperatures but becomes *feasible as the temperature is raised*. Other combinations of ΔH and ΔS give different feasibility conditions, as shown in Table 10.

	ΔS positive (entropy gain)	ΔS negative (entropy loss)
ΔH negative *(exothermic)*	All T	Low T
ΔH positive *(endothermic)*	High T	No T

Table 10
Favourable temperatures for a feasible reaction under different enthalpy and entropy conditions

14.2 Periodicity

The physical properties of the Period 3 elements (Na to Ar) were discussed in *Module 1: Atomic Structure, Bonding and Periodicity*. This section considers their chemical reactions.

14.2.1 *Reactions of Period 3 elements*

With water

Of the **Period 3 elements**, only sodium, magnesium and chlorine react with water. The reaction of chlorine with water has been studied in *Module 2: Foundation Physical and Inorganic Chemistry*, section 11.5.4. Aluminium, silicon, phosphorus and sulphur do not react with water under normal conditions.

Sodium reacts violently with cold water. A piece of sodium added to cold water fizzes, skates over the surface of the water and becomes molten from the heat of the reaction. Hydrogen is evolved and this may catch fire and burn with a yellow flame (characteristic of sodium). At the end of the reaction, a colourless alkaline solution of sodium hydroxide remains:

$$2Na(s) + 2H_2O(l) \rightarrow 2Na^+(aq) + 2OH^-(aq) + H_2(g)$$

By contrast (see *Module 1: Atomic Structure, Bonding and Periodicity*, section 4.10.3), magnesium reacts only very slowly with cold water but burns in steam when heated to give magnesium oxide and hydrogen:

$$Mg(s) + H_2O(g) \rightarrow MgO(s) + H_2(g)$$

With oxygen

The solid elements (Na to S) in Period 3 all burn in air or oxygen when ignited. Sodium burns with a yellow flame, forming the oxide:

$$2Na(s) + \tfrac{1}{2}O_2(g) \rightarrow Na_2O(s)$$

Magnesium, aluminium, silicon and phosphorus burn when ignited, emitting a very bright white light and white smoke of the oxides.

$$Mg(s) + \tfrac{1}{2}O_2(g) \rightarrow MgO(s)$$

$$2Al(s) + \tfrac{3}{2}O_2(g) \rightarrow Al_2O_3(s)$$

$$Si(s) + O_2(g) \rightarrow SiO_2(s)$$

$$P_4(s) + 5O_2(g) \rightarrow P_4O_{10}(s)$$

These reactions are very exothermic.

If there is a limited supply of oxygen, phosphorus also forms phosphorus(III) oxide, P_4O_6.

Sulphur burns with a blue flame but much less vigorously than the elements above, to form the pungent, colourless gas sulphur dioxide:

$$S(s) + O_2(g) \rightarrow SO_2(g)$$

In an excess of pure oxygen, some sulphur trioxide is also formed.

With chlorine

All of the elements from sodium to phosphorus react with chlorine when heated to form the chlorides:

$$
\begin{aligned}
Na(s) \;+\; \tfrac{1}{2}Cl_2(g) &\;\rightarrow\; NaCl(s) \\
Mg(s) \;+\; Cl_2(g) &\;\rightarrow\; MgCl_2(s) \\
2Al(s) \;+\; 3Cl_2(g) &\;\rightarrow\; 2AlCl_3(s) \\
Si(s) \;+\; 2Cl_2(g) &\;\rightarrow\; SiCl_4(g) \\
P_4(s) \;+\; 5Cl_2(g) &\;\rightarrow\; 4PCl_5(s)
\end{aligned}
$$

If a limited supply of chlorine is used, phosphorus forms phosphorus(III) chloride, PCl_3, which is a liquid.

14.2.2 Acid–base properties of the oxides of Period 3 elements

Physical properties, structure and bonding

The melting points, T_m, of the oxides are summarised in Table 11.

Table 11
Period 3 oxides

	Na$_2$O	**MgO**	**Al$_2$O$_3$**	**SiO$_2$**	**P$_4$O$_{10}$**	**SO$_2$**
T_m/K	1548	3125	2345	1883	573	200
Bonding	ionic	ionic	ionic-covalent	covalent	covalent	covalent
Structure	lattice	lattice	lattice	macro-molecular	molecular	molecular

Ionic lattices and macromolecular solids involve strong intermolecular forces and have high melting points. Molecular solids involve weak intermolecular dipole–dipole or van der Waals forces and have low melting points.

The reactions of the oxides with water and their structure and bonding are summarised in Table 12.

Table 12
Reactions of Period 3 oxides with water and approximate pH values of the resulting solutions

Reaction with water	**pH**	**Structure and bonding in the oxide**
$Na_2O(s) + H_2O(l) \rightarrow 2Na^+(aq) + 2OH^-(aq)$	14	ionic lattice
$MgO(s) + H_2O(l) \rightarrow Mg^{2+}(aq) + 2OH^-(aq)$ sparingly soluble	9	ionic lattice
no reaction, Al_2O_3 insoluble	7	ionic–covalent lattice
no reaction, SiO_2 insoluble	7	macromolecular covalent
$P_4O_{10} + 6H_2O \rightarrow 4H_3PO_4$ very soluble; violent reaction; H_3PO_4 is a strong acid	0	molecular covalent
$SO_2 + H_2O \rightarrow H_2SO_3$ moderately soluble; H_2SO_3 is a weak acid	3	molecular covalent
$SO_3 + H_2O \rightarrow H_2SO_4$ very soluble; violent reaction; H_2SO_4 is a strong acid	0	molecular covalent

The trend across the period is:

alkaline oxides ⟶ acidic oxides

Across the period, as the bonding in the oxide changes from ionic to molecular, the solutions of the oxides in water change from alkaline to acidic.

14.2.3 *Reactions of the chlorides of Period 3 elements with water*

As with the oxides, the chlorides of sodium and magnesium are ionic solids with high melting points. Aluminium chloride, however, has covalent character and is a low-melting solid which can be sublimed. Covalent character increases with the non-metals silicon and phosphorus. Silicon tetrachloride is a colourless liquid. Phosphorus pentachloride exists as a covalent molecule in the vapour state but is a solid at room temperature with ions PCl_4^+ and PCl_6^-. This unusual situation is a consequence of the great Lewis acidity (see section 14.5.1) of the PCl_5 molecule which is able to accept a chloride ion from another PCl_5 molecule.

The relationship between the melting points of these chlorides and their bonding and structure is summarised in Table 13.

	NaCl	MgCl$_2$	AlCl$_3$	SiCl$_4$	PCl$_5$
T_m/K	1074	987	450 (sub.)	203	435 (sub.)
Bonding	ionic	ionic	covalent	covalent	covalent
Structure	lattice	lattice	molecular	molecular	molecular

The structure of aluminium chloride was discussed in *Module 1: Foundation Physical and Inorganic Chemistry*, section 10.3.2. It exists as a solid at room temperature. The chlorides of silicon and sulphur are liquids at room temperature.

As Period 3 is crossed, the reactions of the chlorides with water become increasingly more violent. The ionic solids NaCl and MgCl$_2$ dissolve in water to give neutral solutions containing aquated metal ions and aquated chloride ions. Aluminium chloride reacts vigorously with water; the reaction generates much heat and steam is evolved. A weakly acidic solution is formed because the aluminium aqua-ion undergoes the hydrolysis reaction typical of all metal(III) ions (see section 14.5.3). Silicon tetrachloride also reacts vigorously with water, but because silicon(IV) has a higher charge to size ratio than aluminium(III), complete hydrolysis occurs. A solution of the strong acid HCl is formed and a white precipitate of Si(OH)$_4$ appears.

In the vapour phase, PCl$_5$ molecules are this shape:

Work out the shapes of PCl_4^+ and PCl_6^-.

Table 13
Period 3 chlorides

(sub.) means that the compound *sublimes* before melting.

Si(OH)$_4$ forms a polymeric solid sometimes called silicic acid.

Phosphorus pentachloride reacts violently with water forming an acidic solution of the strong acids HCl and H_3PO_4. The equations for all these reactions are given in Table 14.

Table 14
Reactions of Period 3 chlorides with water and approximate pH values of the resulting solutions

Reaction with water	pH	Structure and bonding of the chloride
$NaCl(s) \rightarrow Na^+(aq) + Cl^-(aq)$ The sodium chloride dissolves in water	7	ionic lattice
$MgCl_2(s) \rightarrow Mg^{2+}(aq) + 2Cl^-(aq)$ The magnesium chloride dissolves in water	7	ionic lattice
$AlCl_3(s) \rightarrow Al^{3+}(aq) + 3Cl^-(aq)$ The hydrated aluminium ion is partly hydrolysed: $[Al(H_2O)_6]^{3+}(aq) \rightleftharpoons [Al(H_2O)_5(OH)]^{2+}(aq) + H^+(aq)$	3	molecular covalent
$SiCl_4(l) + 4H_2O(l) \rightarrow Si(OH)_4(s) + 4H^+(aq) + 4Cl^-(aq)$ The silicon tetrachloride is hydrolysed	0	molecular covalent
$PCl_5(s) + 4H_2O(l) \rightarrow H_3PO_4(aq) + 5H^+(aq) + 5Cl^-(aq)$ The phosphorus pentachloride is hydrolysed	0	covalent

E Across the period, the trend in pH of chloride solutions is:

neutral → acidic

Ionic chlorides usually dissolve in water to form neutral solutions containing the parent ions which, in water, are hydrated.

Covalent chlorides are *hydrolysed* by water to form acidic solutions containing HCl.

14.3 Redox equilibria

14.3.1 *Variable oxidation state*

In *Module 2: Foundation Physical and Inorganic Chemistry*, section 11.4.3, redox reactions were identified using the oxidation states of the elements involved and also the definitions:

D Oxidation is the process of electron loss.
Reduction is the process of electron gain.

These general definitions of oxidation and reduction can also be applied to reactions involving elements and ions in the d block of the Periodic Table.

The ability of transition elements to form compounds in which the element is in different oxidation states is one of the most important characteristics of transition elements. This characteristic is central to the behaviour of these elements.

The oxidation state of the transition metal atom in a complex ion is equal to the charge that the element would have if it were a simple ion and not co-ordinated or bonded to other species. The rules which were used to assign oxidation states to elements in s and p blocks of the Periodic Table also apply to d block elements. These rules are given in Table 15.

Species	Oxidation state
Uncombined elements	0
Combined oxygen	−2
Combined hydrogen, except in metal hydrides	+1
Combined hydrogen in metal hydrides	−1
Group I metals in compounds	+1
Group II metals in compounds	+2

Table 15
Rules for assignment of oxidation state

Calculation of oxidation state of d block elements

Example 12

Determine the oxidation state of chromium in the complex ion $Cr_2O_7^{2-}$.

As the overall charge is −2, it can be deduced that:
$(2 \times$ oxidation state of chromium$) + (7 \times$ oxidation state of oxygen$) = -2$

Hence
$(2 \times$ oxidation state of chromium$) - 14 = -2$

Thus
$2 \times$ oxidation state of chromium $= +12$
so each chromium in this complex has an oxidation state of +6.

Other complex ions

Transition elements form complex ions with a wide variety of different ligands. When determining the oxidation state of the central transition metal atom in a complex, it is usually far easier to use the overall charge on the ligand rather than the oxidation state of each atom in the ligand. Some common ligands, together with their overall charges, are given in Table 16.

Name	Ligand	Overall charge
Water	H_2O	0
Ammonia	NH_3	0
Hydroxide	OH^-	−1
Chloride	Cl^-	−1
Cyanide	CN^-	−1
Ethane-1,2-diamine	$NH_2CH_2CH_2NH_2$	0
Ethanedioate	$C_2O_4^{2-}$	−2
Thiosulphate	$S_2O_3^{2-}$	−2
Bis[di(carboxymethyl)amino]ethane	$EDTA^{4-}$	−4

Table 16
Common ligands

The oxidation state of the metal in a complex ion can be worked out by using the same method as in Example 12 above, as shown in Examples 13 and 14.

Example 13

Determine the oxidation state of silver in the complex ion $[Ag(S_2O_3)_2]^{3-}$.

As the overall charge is –3, it can be deduced that:
oxidation state of Ag + ($2 \times$ charge on $S_2O_3^{2-}$) = –3

Hence, using the data from Table 16
oxidation state of Ag – 4 = –3
Hence, Ag has an oxidation state of +1, i.e. Ag(I)

Example 14

Determine the oxidation state of chromium in the complex ion $[CrCl_2(H_2O)_4]^+$.

As the overall charge is +1, it can be deduced that:
oxidation state of Cr + ($2 \times$ charge on Cl^-) + ($4 \times$ charge on H_2O) = +1
Table 16 shows that Cl has a charge of –1 and that H_2O has zero charge.

Hence, oxidation state of Cr – 2 = +1

Hence, Cr has an oxidation state of +3, i.e. Cr(III)

The construction of half-equations for reactions

When constructing half-equations, the following points must be observed:

- only *one* element in a half-equation changes oxidation state
- the half-equation must balance for atoms
- the half-equation must balance for charge

For reactions occurring in aqueous solution, it is also helpful to know that, when constructing a half-equation, it can be assumed that water provides a source of oxygen and that any 'surplus' oxygen is converted into water by reaction with hydrogen ions from an acid. Applying these rules, the half-equation for any redox process can be deduced using one of two alternative methods.

Example 15

Deduce the half-equation for the reduction of VO_3^- to V^{2+} in acid solution.

Method 1: Initial use of oxidation states
In this reaction the oxidation state of vanadium has changed from oxidation state V(V) to oxidation state V(II) and hence vanadium has been reduced. The number of electrons required for this reduction are deduced first:

$$V(V) + 3e^- \text{ forms } V(II)$$

But, as V(V) actually exists as VO_3^-, an acidic solution is needed to supply six hydrogen ions to combine with the three oxygen atoms to form three water molecules. The overall equation is:

$$VO_3^- + 6H^+ + 3e^- \rightarrow V^{2+} + 3H_2O$$

Finally, the overall charge on each side of the equation can be usesd to check that this equation is correct.

Method 2: First balancing for atoms

In this reaction VO_3^- changes to V^{2+}.

To balance the equation for atoms, the three oxygen atoms in oxidation state –2 must be combined with six hydrogen ions, provided by added acid, to form three molecules of water:

$$VO_3^- + 6H^+ \text{ forms } V^{2+} + 3H_2O$$

This equation now balances for atoms but not for charge, with a total charge of +5 on the left-hand side and only +2 on the right-hand side. For balance, three electrons have to be added to the left-hand side to give the half-equation:

$$VO_3^- + 6H^+ + 3e^- \rightarrow V^{2+} + 3H_2O$$

Finally, the change in oxidation states of vanadium can be used to check that this equation is correct.

The construction of overall equations for redox reactions

The overall equation for any redox reaction can be obtained by adding together two half-equations so that the number of electrons given by the reducing agent exactly balances the number of electrons accepted by the oxidising agent.

If the same species appears on both sides of an overall equation obtained by adding two half-equations, it is necessary to cancel this species to give a simpler equation. Water molecules and hydrogen ions are the most common species which need to be treated in this way.

Example 16

In acidic solution, dichromate(VI) ions, $Cr_2O_7^{2-}$, can be reduced to chromium(III), Cr^{3+}, by sulphite ions, SO_3^{2-}, which are oxidised to sulphate ions, SO_4^{2-}. Derive half-equations for the oxidation of sulphite ions to sulphate ions and for the reduction of dichromate(VI) ions to chromium(III). Use these half-equations to derive an equation for the overall reaction.

In the half-equation for the reduction of chromium from oxidation state +6 to +3, an acidic solution is required and the 'surplus' oxygens form water.

$$Cr_2O_7^{2-} + 14H^+ + 6e^- \rightarrow 2Cr^{3+} + 7H_2O$$

In the half-equation for the oxidation of sulphur from oxidation state +4 to oxidation state +6, the additional oxygen has been 'supplied' by a water molecule:

$$SO_3^{2-} + H_2O \rightarrow SO_4^{2-} + 2H^+ + 2e^-$$

The first half-equation shows that six electrons must be supplied for each dichromate(VI) ion reduced. Since the oxidation of each sulphite ion only provides two electrons, then three of these oxidation half-reactions are needed for each dichromate(VI) ion reduction half-reaction. The overall equation for the redox reaction is obtained by addition.

Example 16 (continued)

$$Cr_2O_7^{2-} + 14H^+ + 6e^- \rightarrow 2Cr^{3+} + 7H_2O$$

$$3SO_3^{2-} + 3H_2O \rightarrow 3SO_4^{2-} + 6H^+ + 6e^-$$

$$Cr_2O_7^{2-} + 14H^+ + 3SO_3^{2-} + 3H_2O \rightarrow 2Cr^{3+} + 7H_2O + 3SO_4^{2-} + 6H^+$$

In this example it is necessary to simplify the equation by cancelling three water molecules and six hydrogen ions to give the overall equation:

$$Cr_2O_7^{2-} + 8H^+ + 3SO_3^{2-} \rightarrow 2Cr^{3+} + 4H_2O + 3SO_4^{2-}$$

14.3.2 *Electrode potentials*

Half-equations, electron transfer, reduction and oxidation

E The reaction represented by a *half-equation* is called a *half-reaction*.

E Remember: *adding* electrons *reduces* positive charge.

Half-equations have already been discussed in *Module 2: Foundation Physical and Inorganic Chemistry*, section 11.4.3, and also in section 14.3.1 above. Electron transfer leads to electron gain by one species and to electron loss by another. Electron gain is called **reduction**, electron loss is called **oxidation**. All redox reactions can be expressed as the sum of two half-reactions, one involving electron gain, the other electron loss.

Half-equation convention

By convention, redox half-equations are written as *reductions* (electron addition). This is the convention of the *International Union of Pure and Applied Chemistry* (IUPAC). Two half-equations written conventionally have to be combined into an overall equation by reversing one of them, adjusting electrons in both half-equations to balance, and *adding* (or, if the second half-equation is also written as a *reduction*, by *subtracting*).

For example, the redox reaction between copper(II) ions and metallic zinc can be deduced by combining two half-equations as shown below:

reduction of Cu^{2+}:	$Cu^{2+}(aq) + 2e^- \rightarrow Cu(s)$
oxidation of Zn:	$Zn(s) \rightarrow Zn^{2+}(aq) + 2e^-$
overall (Red + Ox):	$Cu^{2+}(aq) + Zn(s) \rightarrow Cu(s) + Zn^{2+}(aq)$

An example of combination by *subtraction* is given in the section on *Cell reactions* on page 30.

Electrochemical cells

Redox reactions can be studied electrically using an **electrochemical cell**. A cell contains two **electrodes** (metallic conductors) immersed in an **electrolyte** (an ionic conductor) either as an aqueous solution or as a molten salt. An electrode together with its associated electrolyte form an **electrode compartment**. Sometimes both electrodes share the same compartment (Fig 3) but if two compartments with different electrolytes are used, then these are joined by a **salt bridge** (Fig 4).

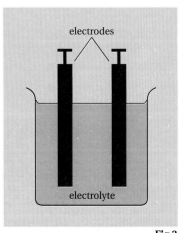

electrodes

electrolyte

Fig 3
A cell with a shared common electrolyte and a single compartment.

A salt bridge consists of an electrolyte solution (often a saturated solution of KCl or KNO_3 in agar jelly) which completes the electrical circuit. It enables the cell to work by allowing **ions** to move between the two compartments while keeping apart the different solutions in the two compartments. If a wire were to be used instead of a salt bridge, it would introduce two more electrodes into the circuit.

Electrode reactions

In an electrochemical cell, each electrode compartment supports its own half-reaction. Electrons released by the oxidation half-reaction in one compartment, e.g.

$$Zn(s) \rightarrow Zn^{2+}(aq) + 2e^-$$

are made available to drive the reduction half-reaction in the other compartment, e.g.

$$Cu^{2+}(aq) + 2e^- \rightarrow Cu(s)$$

by flowing through the external circuit which completes electrical contact between the two compartments.

Electrodes

The simplest electrode of all is the **metal electrode**, which comprises a metal in equilibrium with a solution of its ions. An example is the copper electrode which is written conventionally either as $Cu(s) \mid Cu^{2+}(aq)$ or as $Cu^{2+}(aq) \mid Cu(s)$, depending on the position it occupies in the cell. The vertical bar $\mid$ denotes a boundary between different phases which are themselves specified using state symbols.

The redox couple (e.g. Cu^{2+}/Cu) is a shorthand way of writing a reduction equation (e.g. $Cu^{2+}(aq) + 2e^- \rightarrow Cu(s)$). By convention, the oxidised species (without state symbols) is written first, separated from the reduced species by a forward slash (i.e. *Ox/Red*). This notation is not used in cell diagrams (see page 30) but is an economical way of specifying the half-equation in question.

A **gas electrode** consists of an inert metal (usually platinum) surrounded by a gas in equilibrium with a solution of its ions. The inert metal simply acts as either a source or a sink for electrons. **The hydrogen electrode**, whose special significance will be discussed shortly, is shown schematically in Fig 5 on page 31.

The hydrogen electrode corresponds to the redox couple H^+/H_2 and the electrode is denoted conventionally as

Left $Pt(s) \mid H_2(g) \mid H^+(aq)$ *or* $H^+(aq) \mid H_2(g) \mid Pt(s)$ *Right*

with *two* phase boundaries (solid/gas and gas/liquid) shown as vertical bars. The reduction half-equation at this electrode is:

$$2H^+(aq) + 2e^- \rightarrow H_2(g)$$

A **redox electrode** is one at which two oxidation states of a given element undergo a reduction reaction at an inert metal surface, as in the case of the Fe^{3+}/Fe^{2+} couple

Left $Pt(s) \mid Fe^{2+}(aq), Fe^{3+}(aq)$ *or* $Fe^{3+}(aq), Fe^{2+}(aq) \mid Pt(s)$ *Right*

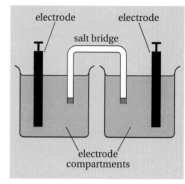

Fig 4
A cell with two electrolytes in separate compartments connected by a salt bridge.

> **E** The electrode at which oxidation occurs is the *negative* electrode; that at which reduction occurs is the *positive* electrode.

> **E** By convention, **cell diagrams** (see below) are always written with metal electrodes on the *outside*. For the electrode on the right, the electrode components are written in the same order as the couple (Ox/Red) with the reverse order (Red/Ox) used for the electrode on the left. Reading through a cell from left to right the order is Red/Ox/Ox/Red or ROOR.

which has a reduction half-reaction, and corresponding half-equation, written as:

$$Fe^{3+}(aq) + e^- \rightarrow Fe^{2+}(aq)$$

The redox couple MnO_4^-, H^+/Mn^{2+} would be denoted as an electrode by

(L) $Pt(s)|Mn^{2+}(aq), MnO_4^-(aq), H^+(aq)$ *or*

$MnO_4^-(aq), H^+(aq), Mn^{2+}(aq)|Pt(s)$ (R)

corresponding to the reduction half-equation:

$$MnO_4^-(aq) + 8H^+(aq) + 5e^- \rightarrow Mn^{2+}(aq) + 4H_2O(l)$$

Cell reactions and spontaneity (feasibility)

There is a simple convention which helps when drawing a **cell diagram** with two electrode compartments joined together to form a cell. The spontaneous cell reaction, in the direction $R \rightarrow L$, will occur if the positive electrode (the one at which reduction occurs) is shown as the right-hand electrode.

The cell diagram is constructed by writing two electrodes back to back, and joining them with a salt bridge, conventionally denoted by two vertical bars:

$$Zn(s) \,|\, Zn^{2+}(aq) \,\|\, Cu^{2+}(aq) \,|\, Cu(s)$$

The cell reaction which corresponds to a given cell diagram can be derived as follows:

- write the right-hand half-equation as a reduction

- beneath it, write the left-hand half-equation as a reduction

- subtract the left-hand half-equation from the right-hand one

For the zinc/copper cell above, the procedure is:

Right half-equation: $Cu^{2+}(aq) + 2e^- \rightarrow Cu(s)$

Left half-equation: $Zn^{2+}(aq) + 2e^- \rightarrow Zn(s)$

overall (R – L): $Cu^{2+}(aq) + Zn(s) \rightarrow Cu(s) + Zn^{2+}(aq)$

This is the spontaneous reaction; zinc does displace copper from solution. This conclusion was the outcome also of the *addition method* in the section on the *Half-equation convention* on page 28.

Of course, the questions that immediately arise are: 'how is it possible to know in advance which of the half-reactions should be on the right (or left)?' or 'which half-reaction will result in reduction (or oxidation)?'

The answer to this question is that it does not matter initially. The choice does not have to be made in advance. As will become clear shortly, the choice of left and right will, of necessity, lead to a **cell potential** (see below) that is either positive or negative. If it is *positive*, the reaction (*R minus L*) will proceed spontaneously as written; if it is *negative*, the reverse reaction will be spontaneous (*feasible*).

E The double bar signifies a junction across which here is no potential difference, usually a salt bridge. Sometimes it is shown as two vertical dashed bars.

E Remember: *R minus L*

Cell potential

In an electrochemical cell, a potential difference (voltage) is set up between the two electrodes. By convention, the left-hand electrode is more negative than the right-hand electrode, which is more positive. The potential difference between the electrodes is called the **cell potential**. Each electrode takes up its own characteristic potential and the overall cell potential is just the difference between the two.

Factors which affect the cell potential

- **Cell current:** the true cell potential can be measured only under **zero-current** conditions, most commonly using a high-resistance digital voltmeter. The cell potential measured with zero current is called the **electromotive force (e.m.f.)**.

- **Cell concentration:** solution concentrations affect the cell potential. The **standard concentration** chosen is 1.0 mol dm^{-3}.

- **Cell temperature:** temperature affects the cell potential. The **standard temperature** chosen is 298 K.

- **Cell pressure:** pressure affects the cell potential, but not significantly unless a gas electrode is used. The **standard pressure** chosen is 100 kPa (1 bar) exactly.

> **E** *Electromotive force (e.m.f.) is the cell potential measured under zero-current conditions. If current is being drawn from the cell, the potential drops.*

The standard cell potential, $E^{\ominus}_{\text{cell}}$

Under standard conditions (zero-current, 1.0 mol dm^{-3}, 298 K, 100 kPa), cell potentials are called **standard potentials** and given the symbol $E^{\ominus}_{\text{cell}}$.

Standard Hydrogen Electrode (SHE)

It is impossible to measure the potential of a single electrode on its own, but if an agreed electrode is assigned the value zero, then all other electrode potentials can be listed relative to this standard. The standard chosen is the hydrogen electrode operating under standard conditions.

It is called the **standard hydrogen electrode** (SHE).

$$\text{Pt(s)} \mid \text{H}_2(\text{g, 1 bar}) \mid \text{H}^+(\text{aq, 1.0 mol dm}^{-3}) \quad E^{\ominus} = 0 \text{ at 298 K}$$

The basic form of this electrode was described earlier (see gas electrodes, page 29) and is shown schematically in Fig 5. Under standard conditions, $E^{\ominus}$(SHE) = 0, by definition.

Standard electrode potential

The **standard electrode potential** of any redox system or couple is found by measuring the potential of a cell with the SHE as the left-hand electrode and the unknown as the right-hand one. However, the hydrogen electrode is very tricky to set up, cumbersome and hazardous, so it is common to use simpler **secondary standards** (see below).

All the variables affecting cell potential listed above *must* be kept under standard conditions when determining standard potentials.

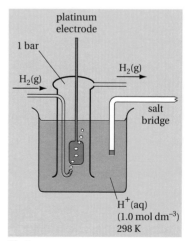

Fig 5
The hydrogen electrode. Hydrogen gas is bubbled over a platinum electrode, establishing an equilibrium with H$^+$(aq).

Calculating $E^{\ominus}$ for an electrode from a measured $E^{\ominus}_{cell}$

The overall cell reaction is determined by subtracting the left-hand half-equation from the right-hand half-equation. The standard cell potential is determined by subtracting the left-hand (L) electrode potential from the right-hand (R) one. Thus

Remember: potentials R – L

$$E^{\ominus}(cell) = E^{\ominus}(R) - E^{\ominus}(L)$$

If the left-hand electrode is the SHE with $E^{\ominus} = 0$, then $E^{\ominus}_{right} = E^{\ominus}_{cell}$

The right-hand electrode is more positive than the left-hand one for a feasible cell reaction.

To ensure that the cell reaction is the spontaneous (*feasible*) one, the less positive standard potential is put to the left and is subtracted from the right hand one, giving $E^{\ominus} > 0$.

Example 17

Calculate the standard e.m.f. of a cell with silver (Ag^+/Ag, $E^{\ominus} = +0.80\ V$) and copper (Cu^{2+}/Cu, $E^{\ominus} = +0.34\ V$) electrodes, each in its own solution of silver(I) ions or copper(II) ions at a concentration of 1.0 mol dm^{-3}, connected by a salt bridge.

Method

The more positive potential (Ag^+/Ag) is made the *right-hand electrode.*

Answer

$E^{\ominus}_{cell} = E^{\ominus}_{right} - E^{\ominus}_{left}$

Therefore

$E^{\ominus}_{cell} = +0.80 - (+0.34) = +0.46\ V$

Example 18

Calculate the standard electrode potential of an unknown metal M which, in a standard solution of its ions $M^{n+}(aq)$, gives a cell potential of +0.48 V as the right-hand electrode of a standard cell with a tin electrode (Sn^{2+}/Sn, $E^{\ominus} = -0.14\ V$) as the left-hand electrode.

Method

Use $E^{\ominus}_{cell} = E^{\ominus}_{right} - E^{\ominus}_{left}$

Answer

M^{n+}/M is the right-hand electrode and $E^{\ominus}_{right} = E^{\ominus}_{cell} + E^{\ominus}_{left}$

$E^{\ominus}(M^{n+}/M) = +0.48 + (-0.14) = +0.34\ V$

Secondary standards

Because the SHE is complicated to set up and use, a more convenient secondary standard electrode, already calibrated against the SHE, is used instead. So, an equally valid determination of an unknown standard potential may be made by measuring relative to a **secondary standard** electrode.

Examples of secondary standards are the *silver/silver chloride electrode* and the *calomel electrode*. Details of the construction of these secondary standard electrodes need not be memorised; neither do their standard potentials, which are:

Silver/silver chloride electrode $\qquad E^{\ominus} = +0.22\text{V}$

Calomel electrode $\qquad E^{\ominus} = +0.27\text{ V}$

The most commonly used secondary electrode is the *calomel electrode*.

Example 19

A standard cell has an e.m.f. of +0.53 V with calomel as the left-hand electrode and the redox couple X^{n+}/X as the right-hand one. Calculate the standard potential of the redox couple X^{n+}/X.

Method

Use $E^{\ominus}_{\text{cell}} = E^{\ominus}_{\text{right}} - E^{\ominus}_{\text{left}}$

Answer

X^{n+}/X is the right-hand electrode and $E^{\ominus}_{\text{right}} = E^{\ominus}_{\text{cell}} + E^{\ominus}_{\text{left}}$

$E^{\ominus}(X^{n+}/X) = +0.53 + 0.27 = +0.80\text{ V}$

Example 20

Calculate the e.m.f. of a cell with the silver/silver chloride electrode ($E^{\ominus} = +0.22$ V) as the left-hand electrode and the Fe^{2+}/Fe electrode ($E^{\ominus} = -0.44$ V) as the right-hand one.

Method

Use $E^{\ominus}_{\text{cell}} = E^{\ominus}_{\text{right}} - E^{\ominus}_{\text{left}}$

Answer

$E^{\ominus}_{\text{cell}} = -0.44 - 0.22 = -0.66\text{ V}$

An *insoluble-salt electrode* consists of a metal covered by a porous layer of an insoluble salt.

The silver/silver chloride electrode is:

$Ag(s) \mid AgCl(s) \mid Cl^-(aq)$

with the reduction half-equation:

$AgCl(s) + e^- \rightarrow Ag(s) + Cl^-(aq)$.

The calomel electrode is

$Pt(s) \mid Hg(l) \mid Hg_2Cl_2(s), KCl(aq)$

with the reduction half-equation

$\frac{1}{2}Hg_2Cl_2(s) + e^- \rightarrow Hg(l) + Cl^-(aq)$

E

The cell e.m.f. is negative, so the cell reaction will *not* be spontaneous (feasible) in the direction R → L. Instead, the reverse reaction occurs:

$2AgCl(s) + Fe(s) \rightarrow$
$\qquad 2Ag(s) + Fe^{2+}(aq) + 2Cl^-(aq)$

14.3.3 *Electrochemical series*

Standard electrode potentials measured relative to the SHE can be placed in a list, either descending in voltage from the most positive (the convention adopted here) or descending from the most negative. The choice is arbitrary, and it is necessary to accept and understand data presented in either format.

The data in Table 17 form a part of the **electrochemical series**, with reduction half-reactions in this table listed in order of decreasing electrode potential. In each case, the reactions listed can be thought of as an oxidising agent accepting electrons to form a reducing agent.

- The strongest *oxidising* agents accept electrons easily and have more positive potentials.

- The strongest *reducing* agents lose electrons easily and have more negative potentials.

- Half-reactions with more positive potentials correspond to *electron gain* (**reduction**) reactions and go readily from left to right.

- Half-reactions with more negative potentials correspond to *electron loss* (**oxidation**) reactions and go readily from right to left.

E

Many of the helpful concepts for interpreting and using the electrochemical series, including those used in this book, depend on the order in which the data are presented. Flexibility of approach is needed to cope with any order of presentation.

Table 17
Standard electrode potentials
at 298 K

E The bromine reduction
half-equation could equally
well be written as:

$\frac{1}{2}Br_2(l) + e^- \to Br^-(aq)$

since it is *potentials* rather than
amounts that are being
considered.

Reduction half-reaction	ox + e⁻ → red	$E^\ominus$/V
	$F_2(g) + 2e^- \to 2F^-(aq)$	+2.87
	$MnO_4^{2-}(aq) + 4H^+(aq) + 2e^- \to MnO_2(s) + 2H_2O(l)$	+1.55
	$MnO_4^-(aq) + 8H^+(aq) + 5e^- \to Mn^{2+}(aq) + 4H_2O(l)$	+1.51
	$Cl_2(g) + 2e^- \to 2Cl^-(aq)$	+1.36
	$Cr_2O_7^{2-}(aq) + 14H^+(aq) + 6e^- \to 2Cr^{3+}(aq) + 7H_2O(l)$	+1.33
	$Br_2(l) + 2e^- \to 2Br^-(aq)$	+1.09
	$Ag^+(aq) + e^- \to Ag(s)$	+0.80
	$Fe^{3+}(aq) + e^- \to Fe^{2+}(aq)$	+0.77
	$MnO_4^-(aq) + e^- \to MnO_4^{2-}(aq)$	+0.56
	$I_2(s) + 2e^- \to 2I^-(aq)$	+0.54
	$Cu^{2+}(aq) + 2e^- \to Cu(s)$	+0.34
	$Hg_2Cl_2(aq) + 2e^- \to 2Hg(l) + 2Cl^-(aq)$	+0.27
	$AgCl(s) + e^- \to Ag(s) + Cl^-(aq)$	+0.22
	$2H^+(aq) + 2e^- \to H_2(g)$	defined as 0
	$Pb^{2+}(aq) + 2e^- \to Pb(s)$	−0.13
Increasing	$Sn^{2+}(aq) + 2e^- \to Sn(s)$	−0.14
Oxidising	$V^{3+}(aq) + e^- \to V^{2+}(aq)$	−0.26
Power	$Fe^{2+}(aq) + 2e^- \to Fe(s)$	−0.44
	$Zn^{2+}(aq) + 2e^- \to Zn(s)$	−0.76
	$Al^{3+}(aq) + 3e^- \to Al(s)$	−1.66
	$Mg^{2+}(aq) + 2e^- \to Mg(s)$	−2.36
	$Na^+(aq) + e^- \to Na(s)$	−2.71
	$Ca^{2+}(aq) + 2e^- \to Ca(s)$	−2.87
	$K^+(aq) + e^- \to K(s)$	−2.93
	$Li^+(aq) + e^- \to Li(s)$	−3.05

(Right-hand column in table: *Increasing Reducing Power*)

E Spontaneous cell reactions
require

$E^\ominus > 0$

Spontaneous (*feasible*)
reactions have $\Delta G^\ominus < 0$
(*negative*). Thus, positive cell
potentials must be linked to
negative $\Delta G^\ominus$ and vice-versa.

$-\Delta G^\ominus$ corresponds to $+ E^\ominus$

E Remember: *positive to the right.*

Spontaneous (*feasible*) direction of a reaction

Spontaneous reactions occur in a cell if the right-hand electrode is the site of *reduction*. When this happens, the right–hand electrode is more positive than the left-hand electrode and $E^\ominus_{right} - E^\ominus_{left}$ is necessarily positive.

D A cell reaction is spontaneous if, and only if, $E^\ominus_{cell}$ is positive.

The more positive the cell potential is, the further the cell equilibrium lies to the right.

In terms of the electrochemical series, the half-reaction with the more positive potential *oxidises* the one with the more negative potential. In terms of the reduction equations in Table 17, the spontaneous direction of reaction (left to right or right to left) involving pairs of half-reactions can be summarised as:

Electrode potential, $E^\ominus$	Spontaneous direction
more +ve	Forwards: L → R
more −ve	Backwards: L ← R

The electrochemical series, cell potentials, and spontaneous reactions

Predictions about spontaneous reactions can be made using the half-reactions in a table of standard redox potentials. A series of six simple steps will always lead to the right answer. These are illustrated in Table 18 below, using the Ag^+/Ag and the Zn^{2+}/Zn redox couples as an example.

Table 18
Six steps in calculating cell e.m.f.

Step 1

Consider the two half-reactions:

$$Ag^+(aq) + e^- \rightarrow Ag(s) \quad E^{\ominus} = +0.80 \text{ V}$$
$$Zn^{2+}(aq) + 2e^- \rightarrow Zn(s) \quad E^{\ominus} = -0.76 \text{ V}$$

Step 2

The half-equation with the *more positive* $E^{\ominus}$ value becomes the *positive electrode*; electrons arrive here from the external circuit, so:

$$Ag^+(aq) + e- \rightarrow Ag(s) \quad or \quad Ag^+/Ag$$

The conventional half-reaction goes *forwards*.

Silver ions behave as the *oxidising* agent and are reduced.

Step 3

The half-equation with the *less positive* $E^{\ominus}$ value becomes the *negative electrode*; electrons leave this electrode and enter the external circuit, so:

$$Zn(s) \rightarrow Zn^{2+}(aq) + 2e^- \quad or \quad Zn/Zn^{2+}$$

The conventional half–reaction goes in *reverse*.

Zinc atoms behaves as the *reducing* agent and are oxidised.

Step 4

The *overall equation* is obtained by adding the two new half-reactions (step 2 and step 3) so that *electrons cancel*:

$$2Ag^+(aq) + Zn(s) \rightarrow Ag(s) + Zn^{2+}(aq)$$

Note the need to double the Ag^+/Ag equation so that the electrons can cancel.

Step 5

The **cell representation** is obtained by placing the couple with the **more positive** $E^{\ominus}$ value on the right, with a salt bridge to separate it from the other couple:

$$\ominus \, Zn(s) \,|\, Zn^{2+}(aq) \,\|\, Ag^+(aq) \,|\, Ag(s) \, \oplus$$

The electrode polarities follow the rule *positive on the right*.

Step 6

The **cell e.m.f.** is obtained using $E^{\ominus}_{cell} = E^{\ominus}_{right} - E^{\ominus}_{left}$, so:

$$E^{\ominus}_{cell} = +0.80 - (-0.76) = +1.56 \text{ V}$$

Several of the stages listed in Table 18, particularly steps 1, 2, 3 and 5, can be represented as a diagram. This procedure is shown in Fig 6 and Fig 7 for the reaction between the Ag^+/Ag and the Zn^{2+}/Zn couples. A sketch based on the electrochemical series (Table 17) is drawn and the two selected half-reactions are identified (*step 1, Table 18*). The *upper* (more positive) couple is marked $\oplus$ and the *lower* (less positive) one is marked $\ominus$ (*steps 2 and 3, Table 18*). All other half-equations and potentials in the electrochemical series can be ignored.

$$\begin{array}{lll}
Cl_2(g) + 2e^- & \rightarrow & 2Cl^-(aq) & +1.36 \\
Br_2(l) + 2e^- & \rightarrow & 2Br^-(aq) & +1.09 \\
Ag^+(aq) + e^- & \rightarrow & Ag(s) & +0.80 \\
Fe^{3+}(aq) + e^- & \rightarrow & Fe^{2+}(aq) & +0.77 \\
I_2(s) + 2e^- & \rightarrow & 2I^-(aq) & +0.54 \\
Cu^{2+}(aq) + 2e^- & \rightarrow & Cu(s) & +0.34 \\
Hg_2Cl_2(aq) + 2e^- & \rightarrow & 2Hg(l) + 2Cl^-(aq) & +0.27 \\
AgCl(s) + e^- & \rightarrow & Ag(s) + Cl^-(aq) & +0.22 \\
2H^+(aq) + 2e^- & \rightarrow & H_2(g) & 0\ defined \\
Pb^{2+}(aq) + 2e^- & \rightarrow & Pb(s) & -0.13 \\
Sn^{2+}(aq) + 2e^- & \rightarrow & Sn(s) & -0.14 \\
Fe^{2+}(aq) + 2e^- & \rightarrow & Fe(s) & -0.44 \\
Zn^{2+}(aq) + 2e^- & \rightarrow & Zn(s) & -0.76
\end{array}$$

+0.80V

−0.76 V

$E_{cell}^{\ominus} = +1.56V$

Fig 6
Determining the direction of spontaneous reaction and e.m.f. for a redox couple

This graphical procedure is simplified by sketching the potential horizontally so that it increases from left to right rather than upwards. Using the two chosen half-equations only, the diagram in Fig 6 is turned clockwise to fit the rule *'positive potential to the right'* (*step 5, Table 18*), and the reaction half-equations are omitted.

The result is an outline sketch, as in Fig 7, from which the cell representation can be written by simple inspection (*step 5, Table 18*). The spontaneous (*feasible*) cell reaction and the cell e.m.f. (*steps 4 and 6, Table 18*) can then be determined using the rule *'R – L'*.

Fig 7
Outline sketch to determine the direction of spontaneous reaction and e.m.f. for a redox couple

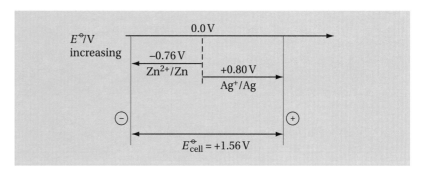

$E^{\ominus}/V$ increasing

0.0 V

−0.76 V
Zn^{2+}/Zn

+0.80 V
Ag^+/Ag

$E_{cell}^{\ominus} = +1.56\ V$

E

Potentials increase left to right.
Right-hand electrode marked +
Reduction at +
Oxidation at −
$E_{cell}^{\ominus} = E_{right}^{\ominus} - E_{left}^{\ominus}$

Note that the half-reaction with the *more positive* potential always goes *forwards* in the spontaneous (feasible) *cell reaction.*

In the example shown, the *cell representation* is:

$^{\ominus}$ $Zn(s) \mid Zn^{2+}(aq) \parallel Ag^+(aq) \mid Ag(s)$ $^{\oplus}$

The spontaneous *cell reaction* is

Positive (Right) electrode:	$2Ag^+(aq) + 2e^- \rightarrow 2Ag(s)$
Negative (Left) electrode:	$Zn^{2+}(aq) + 2e^- \rightarrow Zn(s)$
Overall cell reaction (R – L):	$2Ag^+(aq) + Zn(s) \rightarrow Zn^{2+}(aq) + 2Ag(s)$

and the *cell e.m.f.* is

$$E_{cell}^{\ominus} = E_{right}^{\ominus} - E_{left}^{\ominus} = +0.80 - (-0.76) = +1.56\ V$$

These alternative approaches (*six steps* or *outline sketch*) are entirely equivalent and there is no need to try both. Instead, it is worth choosing the approach which comes most naturally and then practising to gain confidence in its use.

The determination of e.m.f. and overall reaction is illustrated in Examples 21–23 below. In Example 21, both the *'six steps'* and the *'outline sketch'* methods are used, followed by illustrations in Examples 22 and 23.

Example 21

Use data from Table 17 to predict if magnesium will reduce vanadium(III) ions to vanadium(II) ions. Write the representation of a standard cell in which reaction would occur and determine its e.m.f.

Method 1

Carry out the *six steps* in Table 18:

Step 1: $\quad$ $Mg^{2+}(g) + 2e^- \rightarrow Mg(s)$ $\qquad$ $E^{\ominus} = -2.73\,V$
$\qquad\qquad$ $V^{3+}(aq) + 2e^- \rightarrow V^{2+}(aq)$ $\qquad$ $E^{\ominus} = -0.26\,V$

Step 2: $\quad$ $V^{3+}(aq) + 2e^- \rightarrow V^{2+}(aq)$ $\qquad$ $V^{3+}(aq)$ is the oxidising agent

Step 3: $\qquad\qquad$ $Mg(s) \rightarrow Mg^{2+}(aq) + 2e^-$ $\quad$ Mg is *oxidised*

Step 4: $\quad$ $2V^{3+}(aq) + Mg(s) \rightarrow Mg^{2+}(aq) + 2V^{2+}(aq)$

Step 5: $\quad$ $^{\ominus}Mg(s) \mid Mg^{2+}(aq) \mid\mid V^{3+}(aq), V^{2+}(aq) \mid Pt(s)^{+}$

Step 6: $\qquad\qquad$ $E^{\ominus}_{cell} = -0.26 - (-2.37) = +2.11\,V$

Comment

The V^{3+}/V^{2+} couple reacts in the forward direction since it is more positive than the Mg^{2+}/Mg couple. Consequently, Mg will reduce V^{3+} to V^{2+}, itself being oxidised to Mg^{2+}.

Method 2

Draw the *outline sketch* as in Fig 7 above:

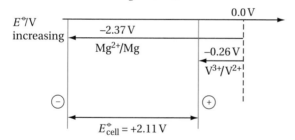

By inspection,
cell representation:

$\quad ^{\ominus}Mg(s) \mid Mg^{2+}(aq) \mid\mid V^{3+}(aq), V^{2+}(aq) \mid Pt(s)^{+}$

cell reaction (R –L):

$\quad Mg(s) + 2V^{3+}(aq) \rightarrow Mg^{2+}(aq) + 2V^{2+}(aq)$

cell e.m.f. (R – L):

$\quad E^{\ominus}_{cell} = -0.26 - (-2.37) = +2.11\,V$

Comment

The V^{3+}/V^{2+} couple reacts in the forward direction since it is more positive than the Mg^{2+}/Mg couple. Consequently, Mg will reduce V^{3+} to V^{2+}, itself being oxidised to Mg^{2+}.

E The *positive electrode* (V^{3+}/V^{2+}) *gains electrons* (from the external circuit)

$$V^{3+} + e^- \rightarrow V^{2+}$$

and the *negative one* (Mg^{2+}/Mg) *loses electrons* (to the external circuit)

$$Mg \rightarrow Mg^{2+} + 2e^-$$

E A platinum electrode is used to make an electrical connection to the V^{3+}/V^{2+} couple. The oxidised and the reduced forms of vanadium are not separated by a phase boundary, so the solid bar is replaced by a comma in Step 5.

The oxidation of iron(II) ions by chlorine is examined in Example 22 using the '*six steps*' method.

Example 22

Use data from Table 17 to determine if chlorine can oxidise Fe(II) ions to Fe(III) ions. Write the representation of a standard cell in which this reaction might occur and determine its e.m.f.

Method 1
Carry out the *six steps* in Table 18:

Step 1:
$$Cl_2(g) + 2e^- \rightarrow 2Cl^-(aq) \qquad E^{\ominus} = +1.36\,V$$
$$Fe^{3+}(aq) + e^- \rightarrow Fe^{2+}(aq) \qquad E^{\ominus} = +0.77\,V$$

Step 2: $\qquad Cl_2(g) + 2e^- \rightarrow 2Cl^-(aq) \qquad$ Cl_2 is the *oxidising* agent

Step 3: $\qquad\qquad Fe^{2+}(g) \rightarrow Fe^{3+}(aq) + e^- \qquad$ Fe^{2+} is *oxidised*

Step 4: $\quad 2Fe^{2+}(aq) + Cl_2(g) \rightarrow 2Fe^{3+}(aq) + 2Cl^-(aq)$

Step 5: $\quad \ominus Pt(s) \mid Fe^{2+}(aq),\, Fe^{3+}(aq) \parallel Cl^-(aq) \mid Cl_2(g) \mid Pt(s) \oplus$

Step 6: $\qquad\qquad E^{\ominus}_{cell} = +1.36 - (+0.77) = +0.59\,V$

Comment
The Cl_2/Cl^- couple reacts in the forward direction since it is more positive than the Fe^{3+}/Fe^{2+} couple. Consequently, Cl_2 will oxidise Fe^{2+} to Fe^{3+}, itself being reduced to Cl^-.

E The *positive electrode* ($Cl_2/2Cl^-$) *gains electrons:*

$$Cl_2 + 2e^- \rightarrow 2Cl^-$$

and the *negative one* (Fe^{3+}/Fe^{2+}) *loses electrons:*

$$Fe^{2+} \rightarrow Fe^{3+} + e^-$$

Finally, in Example 23, the disproportionation of aqueous manganate(VI) ions is examined using the '*outline sketch*' method.

Example 23

Can manganate(VI) ions in aqueous solution disproportionate into manganese(VII) and manganese(IV) species? Write the representation of a standard cell in which this reaction might occur, and determine its e.m.f. Use data from Table 17.

Method 2
Draw the *outline sketch* as in Fig 7 above:

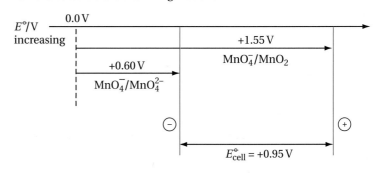

By inspection,
cell representation:

$$\ominus Pt(s) \mid MnO_4^{2-}(aq),\, MnO_4^-(aq) \parallel MnO_4^{2-}(aq) \mid MnO_2(s) \mid Pt(s) \oplus$$

E The *positive electrode* (MnO_4^{2-}/MnO_2) *gains electrons:*

$$4H^+ + MnO_4^{2-} + 2e^- \rightarrow MnO_2 + 2H_2O$$

and the *negative one* (MnO_4^-/MnO_4^{2-}) *loses electrons:*

$$MnO_4^{2-} \rightarrow MnO_4^- + e^-$$

cell reaction (R − L):
$$4H^+(aq) + 3MnO_4^{2-}(aq) \rightarrow 2MnO_4^-(aq) + MnO_2(s) + 2H_2O(l)$$

cell e.m.f. (R − L):
$$E_{cell}^{\ominus} = +1.55 - (+0.60) = +0.95\ V$$

The MnO_4^{2-}/MnO_2 couple reacts as shown in the equation (i.e. in the forward direction) since it is more positive than the MnO_4^-/MnO_4^{2-} couple. Consequently, manganate(VI) ions in aqueous solution can disproportionate into manganate(VII) and manganese(IV) species.

14.4 | Transition metals

14.4.1 *General properties*

The transition elements occupy the large central block of the Periodic Table, the **d-block** (see *Module 1: Atomic Structure, Bonding and Periodicity*, section 10.4.1). However, in the present definition of a transition element, not all d-block elements are considered to be transition elements.

> **A transition element is an element having an incomplete d (or f) shell either in the element or in one of its common ions.** | **D**

The 3d shell can contain up to ten electrons; there are therefore ten elements in the first row of the d-block. The first element is scandium, the electron arrangement of which can be deduced from the Periodic Table as:

Sc: [Ar]$3d^14s^2$

> Note that [Ar]$4s^23d^1$ is equally acceptable as a representation. | **E**

This electron arrangement has an incomplete d-shell so that scandium is a transition element. Across the Periodic Table, the d-shell is progressively filled until copper which has the electron arrangement

Cu: [Ar]$3d^{10}4s^1$

> This electron arrangement is more stable than [Ar]$3d^94s^2$. | **E**

This electron arrangement does not have an incomplete d-shell. However, one of the common ions of copper is Cu^{2+}, which does have a electron arrangement with an incomplete d-shell:

Cu^{2+}:   [Ar]$3d^9$

Copper is therefore a transition element.

After copper, the next element is zinc, which has the electron arrangement:

Zn: [Ar]$3d^{10}4s^2$

and it forms only one common ion which has the electron arrangement

Zn^{2+}:   [Ar]$3d^{10}$

Neither in the element nor in its common ion does zinc have an incomplete d-shell so that zinc is not classed as a transition element.

In working out the electron arrangement of a transition-metal ion, the *outer s-electrons are always lost first* from the electron arrangement of the metal. Transition-metal compounds do not have any outer s-electrons; it is the partly filled d-shell which is responsible for the characteristic properties of the transition-metal ions in their compounds.

Characteristic properties of transition elements are:

- formation of complexes
- formation of coloured ions
- variable oxidation states
- catalytic activity

14.4.2 *Complex formation*

Complex compounds contain a central atom surrounded by ions or molecules, both of which are called ligands.

E Strictly, a ligand is a lone-pair donor only when it is actually bonded to a metal ion.

D *A ligand is any atom, ion or molecule which can donate a pair of electrons to a metal ion.*

The ability to donate a pair of electrons means that:

D *ligand = Lewis base = nucleophile*

When a complex compound is formed, ligands donate an electron pair to a metal ion to form a **co-ordinate bond** with the metal.

E In a complex compound, the co-ordination number of the metal differs from its oxidation state.

D *The number of atoms bonded to the metal ion is called the* **co-ordination number.**

In the hexaaquacopper(II) ion, each water molecule donates an electron pair (one of its lone pairs) to the copper(II) ion to form an octahedral complex:

$$CuSO_4(s) \xrightarrow{\;H_2O\;} [Cu(H_2O)_6]^{2+}$$

copper(II) sulphate   hexaaquacopper(II) ion

white   blue

Co-ordination number of Cu = 6

Oxidation state of Cu = +2

For ligands with only one donor atom, the co-ordination number is the number of ligands bonded to the metal ion; this is not true for multidentate ligands (see below).

It is important to remember that whilst the hexaaqua complex is an ion, the bonds within the complex, i.e. the Cu–O and the O–H bonds are covalent (see section 14.5.2 for a diagram of the hexaaqua ion). Reactions of the complex ion involve the breaking of one or both of these types of bond.

A ligand such as water, which has only one atom that can donate a pair of electrons, and which consequently bonds through one atom only, is said to be **unidentate**. Unidentate ligands include:

H_2O, NH_3, Cl^-, OH^- and CN^-

Although several of these species have more than one lone pair of electrons, each ligand donates only one lone pair on co-ordination.

Ligands which contain two donor atoms, and which consequently are able to bond to a metal ion through two atoms, are called **bidentate**. Bidentate ligands include:

ethane-1,2-diamine (*ethylenediamine* or *en*), $H_2NCH_2CH_2NH_2$ (Fig 8), which bonds through *two nitrogen atoms* and the ethanedioate (*oxalate*) ion, $C_2O_4^{2-}$ (Fig 9), which bonds through *two oxygen atoms*

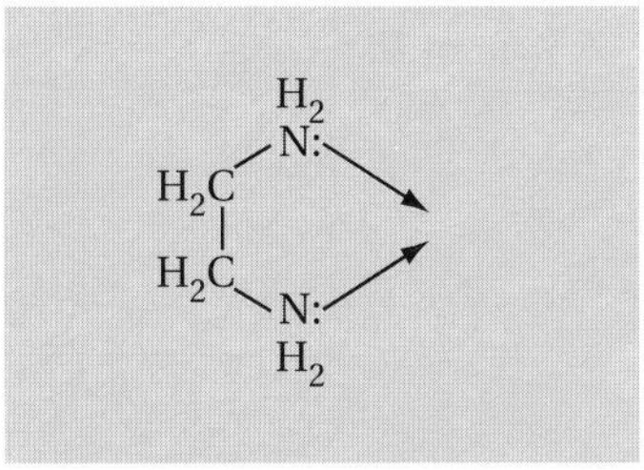

Fig 8
Ethane-1,2-diamine
(*ethylenediamine*).

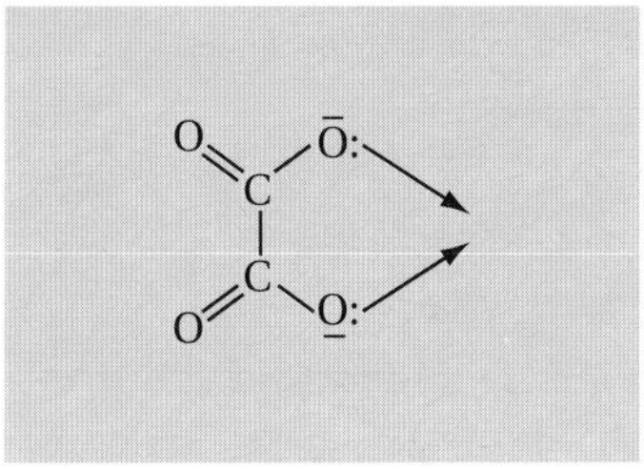

Fig 9
The ethanedioate (*oxalate*) ion.

Some ligands contain many donor atoms and are said to be **multidentate**. Typical of these is the anion derived from bis[di(carboxymethyl)amino]ethane, commonly known as ethylenediaminetetraacetic acid or H_4EDTA. The anion $EDTA^{4-}$ is able to bond to metal ions from six donor atoms.

The structure of the $EDTA^{4-}$ anion is shown in Fig 10.

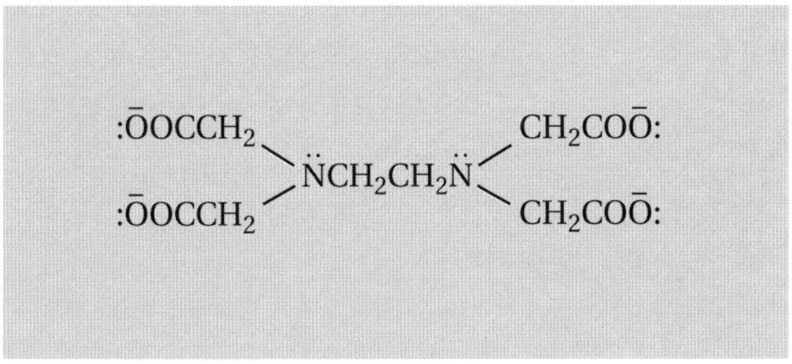

Fig 10
The structure of the $EDTA^{4-}$ anion.

EDTA can be used to estimate many metals volumetrically. An indicator is used which forms a weak complex with the metal ion; when all the free metal ion has been complexed with EDTA, the indicator is displaced from its weak metal complex and a new colour is seen at the end-point.

EDTA uses the lone pairs on its six donor sites (4O and 2N) and forms 1:1 complexes with metal(II) ions, e.g.

$$[Cu(H_2O)_6]^{2+} + EDTA^{4-} \rightarrow [Cu(EDTA)]^{2-} + 6H_2O$$

Blood contains a red iron(II) complex called haem. In this complex, shown in Fig 11, the iron is bonded to four nitrogen atoms in a plane within a large organic molecule called porphyrin (which is therefore a *quadridentate* ligand). Octahedral co-ordination of the iron involves a fifth nitrogen atom, above this plane, from a protein called globin, with the sixth position, below the plane, occupied by either molecular oxygen or the oxygen atom of a water molecule.

Fig 11
An iron(II) ion at the centre of a porphyrin ring in haem. Various organic side chains are attached to the five-membered rings.

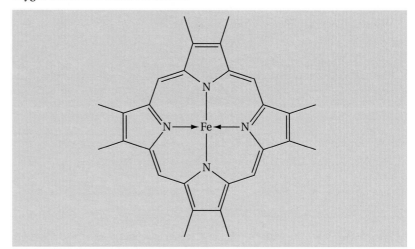

14.4.3 *Shapes of complex ions*

The most common shape of complex ions is the *octahedral* shape formed, for example, when water molecules co-ordinate to a copper(II) ion. All the transition metals of the first-row transition series form octahedral hexaaqua ions with water. Ammonia is another ligand with no charge; it is similar in size to water and readily forms octahedral complexes with these metals, as will be seen later in section 14.5.4.

The next most common shape encountered in complexes is *tetrahedral*. If the ligand is large and negatively charged (so that there is inter-ligand repulsion as well as bond-pair repulsion in the complex), then it may not be possible to fit six ligands around the central metal ion in a stable complex, so the tetrahedral arrangement is preferred. Ligands are further apart in a tetrahedral complex than they are in an octahedral complex. Chloride ions, bromide ions and iodide ions are typical of large anions which form tetrahedral complexes; examples will be seen later in section 14.5.4.

E Fluoride ions do form octahedral complexes, e.g. $[AlF_6]^{3-}$ and $[FeF_6]^{3-}$, because the F^- ion is smaller than the Cl^- ion.

Less common shapes include the *square-planar* shape in, for example, cisplatin (see section 14.4.7) and the *linear* shape commonly found in silver complexes.

The arrangements of the bonds in the different shapes of complexes are shown in Fig 12.

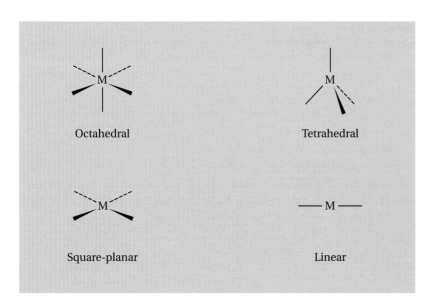

Fig 12
The shapes of complex ions.

Complexes of silver

Octahedral complexes of the transition elements are very common; tetrahedral complexes tend to occur with large and negatively charged ligands. Silver ions are unusual in preferring to form linear complexes. In these, the silver ion is bonded to two ligands only, and the bond angle is 180°. Examples of such complexes are:

$[Ag(NH_3)_2]^+$ i.e. $[H_3N{\rightarrow}Ag{\leftarrow}NH_3]^+$

$[Ag(S_2O_3)_2]^{3-}$ i.e. $[O_3S_2{\rightarrow}Ag{\leftarrow}S_2O_3]^{3-}$

$[Ag(CN)_2]^-$ i.e. $[NC{\rightarrow}Ag{\leftarrow}CN]^-$

Each straight arrow above shows a lone pair of electrons being donated to form a co-ordinate bond. Note the similarity of the straight arrow to the *curly arrow* used in organic chemistry (*Module 3, Introduction to Organic Chemistry*, section 12.3.2). The straight arrow also starts at a pair of electrons and points to where they end up in the new species.

The formation and uses of these complexes will be discussed in section 14.4.7.

> The Ag^+ ion with the outer electron arrangement $4d^{10}$ has no vacant d orbitals; other d^{10} ions such as Cu^+ also give linear shapes and give rise to colourless complexes. **E**

14.4.4 *Formation of coloured ions*

Many of the coloured solutions which are seen around the laboratory owe their colour to the presence of a transition metal. Indeed, along with other evidence, the colour of a solution can be used to identify the presence of a particular transition metal.

When a colour change occurs in the reaction of a transition-metal ion, there is a change in at least one of the following factors:

- oxidation state
- co-ordination number
- ligand

Often there is a change in more than one of these factors.

E See section 14.5.3 for an explanation of the unusual brown colour seen in iron(III) solutions. Similarly the green colour (emerald) usually seen in chromium(III) compounds is due to hydrolysis or substitution by other anions, as in $[Cr(H_2O)_4Cl_2]^+$.

E In some reactions, the colour change arises as a result of all three changes, e.g.

$[CrO_4]^{2-}$ → $[Cr(H_2O)_4Cl_2]^+$
Cr(VI)   Cr(III)
yellow   green
tetrahedral   octahedral

and

$[Mn(H_2O)_6]^{2+}$ → $[MnO_4]^-$
Mn(II)   Mn(VII)
very pale pink purple
octahedral   tetrahedral

$h = 6.63 \times 10^{-34}$ J s

Change of oxidation state

Examples of a change in oxidation state being responsible for the colour change are:

$[Fe^{II}(H_2O)_6]^{2+}$ → $[Fe^{III}(H_2O)_6]^{3+}$
green   very pale violet

$[Cr^{III}(H_2O)_6]^{3+}$ → $[Cr^{II}(H_2O)_6]^{2+}$
red-violet   blue

Change of co-ordination number

A change in co-ordination number (i.e. the number of nearest neighbours) is most usually achieved by changing ligands; examples of colour changes arising from this type of change are:

$[Cu(H_2O)_6]^{2+}$ → $[CuCl_4]^{2-}$
blue   yellow-green
octahedral   tetrahedral

$[Co(H_2O)_6]^{2+}$ → $[CoCl_4]^{2-}$
pink   blue
octahedral   tetrahedral

Change of ligand

Colour changes arising from a change of ligand only are shown by the reactions:

$[Cr(H_2O)_6]^{3+}$ → $[Cr(NH_3)_6]^{3+}$
red-violet   purple
octahedral   octahedral

$[Cu(H_2O)_6]^{2+}$ → $[Cu(NH_3)_4(H_2O)_2]^{2+}$
blue   blue-violet
octahedral   octahedral

Origin of colour

Colour arises when a molecule or an ion absorbs visible light and enters a higher or **excited** energy state. If only a part of the visible spectrum is absorbed, then the eye detects those frequencies which remain and a colour is seen.

Energy is absorbed when an electron is promoted from a lower to a higher energy level, as shown in Fig 13 opposite. The energy difference between the two levels involved is given by

$$E_2 - E_1 = \Delta E = h\nu$$

where h is Planck's constant and ν is the frequency of the absorbed radiation.

If the energy levels are too far apart for visible light to effect the transition, ultraviolet light, which has higher energy, can be used. As the human eye cannot see absorption in the ultraviolet region, it is necessary to use a detector capable of measuring ultraviolet as well as visible radiation.

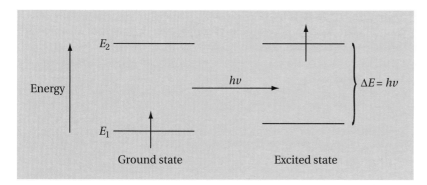

Fig 13
Absorption of light giving rise to colour. The symbol ↑ represents an electron.

If both levels involved are levels of d orbitals, the change is called a '*d–d*' *transition* and the associated colour is not very intense. If, however, the transition involves an electron moving from a ligand orbital to a metal (d) orbital (or vice-versa), the transition is called '*charge-transfer*' and an intense colour results.

Ultraviolet and visible spectrophotometry

The intensity of colours shown by a solution of a transition-metal ion can be used to determine the concentration of that ion; this is done using a spectrophotometer.

In an ultraviolet and visible spectrophotometer, light of increasing (or decreasing) frequency is passed through the sample; the emergent light is received in a detector and recorded. The *amount of light absorbed is proportional to the concentration of the absorbing species* in the solution under test. By measuring how much light is absorbed by a solution at a particular frequency, the concentration of the absorbing species can be determined.

The Beer–Lambert law:
$A = \varepsilon c l$
links absorbance, A, to concentration, c, cell path-length, l, and molar absorption coefficient, ε.

Solutions of transition-metal aqua ions, e.g. the pale green $[Fe(H_2O)_6]^{2+}$ or the pale violet $[Fe(H_2O)_6]^{3+}$, do not absorb very strongly. Consequently, accurate quantitative determination of the concentrations of such solutions is not possible because the difference in absorption between solutions of different concentration is not great enough. The absorbance of such solutions can be increased by a factor of about 10^3 by adding a ligand which gives an intense colour when it co-ordinates to the metal ion. In the case of the iron examples given above, bipyridyl (*bipy*) (see Fig 14) could be used as the ligand for iron(II), and thiocyanate ions as the ligand for iron(III):

$$[Fe(H_2O)_6]^{2+} + 3bipy \rightarrow [Fe(bipy)_3]^{2+} + 6H_2O$$
pale green   intense red

Fig 14
2,2'-Bipyridyl (*bipy*).

$$[Fe(H_2O)_6]^{3+} + NCS^- \rightarrow [Fe(H_2O)_5(NCS)]^{2+} + H_2O$$
pale violet   intense blood-red

The thiocyanate ions, NCS^-, come from an ionic salt such as potassium thiocyanate, $K^+ (NCS^-)$.

The colour of these red complexes of iron arises from *charge-transfer* (metal to ligand in the case of $[Fe(bipy)_3]^{2+}$ and ligand to metal in the case of $[Fe(H_2O)_5(NCS)]^{2+}$; the aqua ion colours arise from *d–d transitions*.

So sensitive are these colours to a change in the concentration of the iron, that a very simple spectrophotometer can be used to measure the intensity of absorption at the particular frequency absorbed. Such spectrophotometers are called **colorimeters** (because they measure the intensity of colours).

14.4.5 *Variable oxidation states*

A characteristic property of the transition elements is their ability to change oxidation state in chemical reactions.

In an oxidation–reduction or 'redox' reaction, the transition-metal ion is oxidised or reduced, i.e. it changes its oxidation state. Because transition-metal ions show variable oxidation states, many redox reactions occur.

Metallic zinc is a good reducing agent and reduces aqueous transition-metal ions to low oxidation states (see section 14.3.3). Zinc reacts with mineral acids to form zinc(II) ions and release two electrons for the reduction:

$$Zn \rightarrow Zn^{2+} + 2e^-$$

Reductions with zinc proceed through the oxidation states of the metal in sequence so that each different colour of the metal ion in every oxidation state can often be seen. Two good examples of this behaviour are seen in the reduction of vanadium(V) and of chromium(VI).

Oxidation states of vanadium

When white ammonium vanadate(V), NH_4VO_3, is added to dilute hydrochloric acid, the orange colour of the dioxovanadium(V) ion, VO_2^+, is seen:

$$VO_3^- + 2H^+ \rightarrow VO_2^+ + H_2O$$
$$\text{white} \qquad\quad \text{orange}$$

E	The ending *-ate* means that the metal is present in a negatively-charged ion (an anion).

E	Note that the change from VO_3^- to VO_2^+ is not a redox reaction.

The addition of granulated zinc to this mixture results in colour changes which occur over several minutes, as the reduction proceeds. The oxidation states, and colours of the ions seen, are:

V(V)	$VO_2^+(aq)$		yellow
↓	↓		↓
V(IV)	$VO^{2+}(aq)$	i.e. $[VO(H_2O)_5]^{2+}$	blue
↓	↓		↓
V(III)	$V^{3+}(aq)$	i.e. $[VCl_2(H_2O)_4]^+$	green
↓	↓		↓
V(II)	$V^{2+}(aq)$	i.e. $[V(H_2O)_6]^{2+}$	violet

If sulphuric acid is used instead of hydrochloric acid in this reduction, the vanadium(III) species seen will have the dull grey-blue colour of the hexaaqua ion $[V(H_2O)_6]^{3+}$. Since vanadium(II) ions are oxidised by air, it may be necessary to carry out this reduction in a lightly stoppered flask.

Oxidation states of chromium

Chromium shows its highest oxidation state of +6 in the yellow chromate(VI) ion, CrO_4^{2-}, and the orange dichromate(VI) ion, $Cr_2O_7^{2-}$. The dichromate ion is formed from the chromate ion in acid solution:

$$2CrO_4^{2-} + 2H^+ \rightarrow Cr_2O_7^{2-} + H_2O$$
yellow   orange

The reduction of dichromate(VI) ions by zinc and hydrochloric acid proceeds through the following stages:

Cr(VI)	$Cr_2O_7^{2-}$		orange
↓	↓		↓
Cr(III)	$Cr^{3+}(aq)$	i.e. $[CrCl_2(H_2O)_4]^+$	green
↓	↓		↓
Cr(II)	$Cr^{2+}(aq)$	i.e. $[Cr(H_2O)_6]^{2+}$	blue

In order to reach the chromium(II) stage, it is necessary to keep air away from the reduction. When air is admitted to the blue chromium(II) solution, the colour rapidly changes to green due to the presence of chromium(III). Once again it is clear that, in hydrochloric acid, some substitution of the chromium(III)-aqua ion has taken place. Even in sulphuric acid, substitution of sulphate ions into the hexaaqua ion occurs so that the red-violet (ruby) colour of the $[Cr(H_2O)_6]^{3+}$ ion is not seen.

Oxidation states of manganese

Manganese, the next element along the first transition series, shows its maximum oxidation state of +7 in the manganate(VII), MnO_4^-, ion. This ion is a powerful oxidising agent which is reduced directly to the manganese +2 state by zinc or by iron or even by iron(II) ions in acid solution.

The $[Mn(H_2O)_6]^{2+}$ ion is a very pale pink colour and appears colourless in aqueous solution so that, when manganate(VII) ions are reduced to Mn(II) ions, the colour change seen is purple to colourless:

Mn(VII)	MnO_4^-	purple
↓	↓	↓
Mn(II)	$[Mn(H_2O)_6]^{2+}$	colourless

Redox titrations

Redox reactions can be used in quantitative volumetric analysis. The end-point, when the reaction is complete, can be determined in a variety of ways, including the use of indicators or electrical methods. The use in redox titrations of two common quantitative oxidising agents is discussed below.

The chromate/dichromate change is not a redox reaction; it is reversed in alkaline solution:
$$Cr_2O_7^{2-} + 2OH^- \rightarrow 2CrO_4^{2-} + H_2O$$
It is an *equilibrium reaction*.

The red-violet colour of hexaaquachromium(III) is seen in solid compounds containing this ion, e.g. 'chrome alum'. This colour is sometimes described as *ruby*; natural gemstones of ruby owe their colour to the octahedral arrangement of oxygen atoms around chromium(III). The gemstone *emerald* also owes its colour to six-coordinate chromium(III) ions but the six ligands are not now identical.

Potassium manganate(VII) titrations

When potassium manganate(VII) is used as an oxidising agent in acidic solution, no additional indicator is required. Because the reduction product, $[Mn(H_2O)_6]^{2+}$, is essentially colourless, no colour is seen until an excess of potassium manganate(VII) has been added, at which point the presence of even a slight excess of the dark purple manganate(VII) ion is readily seen and is used to indicate complete reaction. Knowing the initial and final species present, MnO_4^- and Mn^{2+}, a half-equation can be constructed as follows:

$$MnO_4^- + 8H^+ + 5e^- \rightarrow Mn^{2+} + 4H_2O$$

The presence of an excess of acid for this reaction is essential and the choice of the actual acid used is important. The acid must:

- be strong, because a high concentration of hydrogen ions is needed

- not be an oxidising agent, because an oxidising agent might react with the reducing agent being estimated

- not be a reducing agent, because reducing agents will be oxidised by the manganate(VII) ions

A review of common laboratory acids enables the correct choice of acid to be made:

- it *cannot* be hydrochloric acid, as this can be oxidised to chlorine

- it *cannot* be nitric acid, as this is an oxidising agent

- it *cannot* be *concentrated* sulphuric acid, as this too is an oxidising agent

- it *cannot* be ethanoic acid, as this is a *weak* acid which produces too low a concentration of H^+ ions

Dilute sulphuric acid is, therefore, the main choice among common laboratory acids.

The usual procedure is to add a standard solution of potassium manganate(VII) from a burette to a conical flask containing a solution of the reducing agent which has been acidified by the addition of dilute sulphuric acid. As mentioned above, no additional indicator is required for this titration.

Consider, for example, the use of potassium manganate(VII) to estimate the concentration of iron(II) ions in solution. Knowing that the oxidation of iron(II) to iron(III) is a one-electron change and that the conversion of manganese(VII) into manganese(II) is a five-electron change, it is easy to deduce that the complete reduction of one mole of manganese(VII) to manganese(II) will require five moles of iron(II). The two half-equations and the overall equation for these processes are:

$$MnO_4^- + 8H^+ + 5e^- \rightarrow Mn^{2+} + 4H_2O$$
$$5Fe^{2+} \rightarrow 5Fe^{3+} + 5e^-$$
$$\overline{5Fe^{2+} + MnO_4^- + 8H^+ \rightarrow 5Fe^{3+} + Mn^{2+} + 4H_2O}$$

E If insufficient acid is present, a brown coloration of insoluble MnO_2 is seen.

Potassium dichromate(VI) titrations

For redox titrations using potassium dichromate(VI) as the oxidant, an indicator is required. One of the more commonly used indicators is sodium diphenylaminesulphonate, which turns from colourless to purple at the end-point.

Potassium dichromate(VI) can also be used to estimate iron(II) ions in solution. Because chromium is reduced from the +6 to the +3 oxidation state, the reacting ratio is one mole of chromium(VI) to three moles of iron:

$$Cr_2O_7^{2-} + 14H^+ + 6e^- \rightarrow 2Cr^{3+} + 7H_2O$$

$$6Fe^{3+} \rightarrow 6Fe^{3+} + 6e^-$$

$$\overline{6Fe^{2+} + Cr_2O_7^{2-} + 14H^+ \rightarrow 6Fe^{3+} + 2Cr^{3+} + 7H_2O}$$

Again, such titrations are normally carried out with the dichromate(VI) solution in a burette and the iron(II) solution in a conical flask.

As with manganate(VII) titrations, H^+ is a reactant, so the acid must be present in excess. For titrations involving dichromate(VI), the H^+ ions can be provided either by dilute sulphuric acid or by dilute hydrochloric acid. It is possible to use hydrochloric acid in this case because, unlike the MnO_4^-/Mn^{2+} couple, the potential of the $Cr_2O_7^{2-}/Cr^{3+}$ couple is not positive enough to oxidise chloride ions to chlorine, so that the concentration of iron(II) ions on their own can be measured without interference from chloride ions.

Below are two examples which involve the estimation of iron(II) using each of these volumetric oxidising agents. Example 24 shows the analysis by titration with potassium manganate(VII) solution of a sample of lawn sand which contains iron(II) sulphate.

> Note that there are two **E** moles of chromium(VI) in one mole of dichromate(VI).

Typical laboratory experiments include:

- determining the percentage of iron(II) in commercial lawn sand or iron tablets

- determining the percentage of sulphite ions in a given solid, or in wine to which it has been added as a preservative

---

Example 24

3.00 g of a lawn sand containing an iron(II) salt were shaken with dilute sulphuric acid. The resulting solution required 25.00 cm³ of a 0.0200 mol dm⁻³ solution of KMnO₄ to oxidise all the Fe²⁺ ions in the solution to Fe³⁺ ions. Calculate the percentage by mass of Fe²⁺ ions in the lawn sand.

Method
Calculate the number of moles of manganate(VII) that have reacted, and then use the stoichiometric equation to relate this number to the number of moles of Fe(II) that have been oxidised. Use the relative atomic mass of iron to convert moles into mass and then express the mass as a percentage of the original sample.

Answer
$$\text{moles KMnO}_4 = \frac{0.0200 \times 25.00}{1000} = 5.00 \times 10^{-4} \text{ mol}$$

There are *five* moles of Fe^{2+} for every mole of MnO_4^- used, so

$$\text{moles Fe}^{2+} = 5 \times 5.00 \times 10^{-4} = 2.50 \times 10^{-3} \text{ mol}$$

Example 24 (continued)

Multiplying the number of moles by the mass of one mole of Fe^{2+} gives the mass of Fe^{2+} present in 3.00 g of lawn sand:

mass of Fe^{2+} present $= 55.8 \times 2.5 \times 10^{-3} = 0.140$ g

Hence, the percentage by mass of Fe^{2+} ions present in the sample is

$$\frac{55.8 \times 2.5 \times 10^{-3} \times 100}{3.00} = 4.65\%$$

Comment

Note that it is not necessary to filter off the sand in this analysis because it is inert and does not interfere with the titration.

Note also that *rounding-off* the mass of Fe^{2+} to 3 significant figures *before* calculating the percentage gives a slightly different, and less accurate, answer.

Example 25 shows the analysis by titration with potassium dichromate(VI) solution of a medicinal iron tablet which contains iron(II) sulphate hydrate and an inert filler.

Example 25

When a medicinal iron tablet weighing 0.940 g was dissolved in dilute sulphuric acid and the resulting solution was titrated with a 0.0160 mol dm^{-3} solution of $K_2Cr_2O_7$, exactly 32.5 cm^3 of the potassium dichromate(VI) solution were required to reach the end-point. Calculate the percentage by mass of Fe^{2+} in the tablet.

Method

Calculate the number of moles of dichromate(VI) that have reacted, and then use the stoichiometric equation to relate this number to the number of moles of Fe(II) that have been oxidised. Use the relative atomic mass of iron to convert moles into mass and then express the mass as a percentage of the original tablet.

Answer

$$\text{moles } Cr_2O_7^{2-} = \frac{32.5 \times 0.0160}{1000} = 5.20 \times 10^{-4}$$

There are *six* moles of Fe^{2+} for every mole of $Cr_2O_7^{2-}$ used, so

moles of $Fe^{2+} = 6 \times 5.20 \times 10^{-4} = 3.12 \times 10^{-3}$

Multiplying the number of moles by the mass of one mole of Fe^{2+} gives the mass of Fe^{2+} present in the 0.940 g tablet:

mass of Fe^{2+} is $55.8 \times 6 \times 5.20 \times 10^{-4} = 0.174$ g

Hence, the percentage by mass of Fe^{2+} ions in the tablet is:

$$\frac{55.8 \times 6 \times 5.20 \times 10^{-4} \times 100}{0.940} = 18.5\%$$

Comment
Iron is needed by the body for good health. It is an essential component of haem (see section 14.4.2) which is involved in the transport of oxygen by the blood. A lack of iron leads to the condition known as anaemia. Anaemia sufferers take iron tablets; these tablets are also given to blood donors to restore their iron levels after giving blood.

Oxidation in alkaline solution

In the above titrations, the transition metal in a high oxidation state has been reduced in acid solution. In alkaline solution, transition metal ions in low oxidation states are readily oxidised to higher oxidation states. For example, if a solution of Fe^{2+} ions is to be stored without being oxidised, it must be acidified. In alkaline solution, the precipitated green $Fe(OH)_2$ oxidises rapidly to form brown $Fe(OH)_3$.

Consider the species present:

$[M(H_2O)_6]^{2+}$	$[M(H_2O)_4(OH)_2]$	$[M(OH)_4]^{2-}$
acid solution	neutral	alkaline solution
harder to oxidise		easier to oxidise

Oxidation is electron loss. It is easier to remove an electron from a neutral or negatively-charged species than from a positively-charged one.

To prepare a metal complex which contains the metal in a high oxidation state, the general procedure is:

- first, add an alkali

- then add an oxidising agent

This method is applied in the examples which follow.

The addition of an excess of sodium hydroxide to a solution of a chromium(III) salt gives a green solution containing the green chromate(III) ion (see section 14.5.3). When this solution is treated with hydrogen peroxide, a yellow solution containing chromate(VI) ions is formed. Thus, chromium(III) is readily oxidised to CrO_4^{2-} by hydrogen peroxide in alkaline solution:

$$2[Cr(OH)_6]^{3-} + 3H_2O_2 \rightarrow 2CrO_4^{2-} + 2OH^- + 8H_2O$$

Often, oxygen in the air can act as the oxidising agent; it is capable of oxidising $Fe(OH)_2$ to $Fe(OH)_3$, $Mn(OH)_2$ to $Mn(OH)_3$ and $Co(OH)_2$ to $Co(OH)_3$. Similar oxidations occur when hydrogen peroxide is added. The metal 2+ ions can be stabilised against atmospheric oxidation, however, by keeping them in acidic solution (see storage of iron(II) solutions above).

The addition of aqueous ammonia to a solution of a cobalt(II) salt gives a green-blue precipitate of cobalt(II) hydroxide. If this suspension is shaken in air, oxidation to the brown cobalt(III) hydroxide takes place. If, however, the cobalt(II) hydroxide is treated with an excess of concentrated aqueous ammonia in the absence of air, a pale brown (straw-coloured) solution of the cobalt(II) hexaammine is formed. If air is then admitted, the

Reduction occurs in acid solution, oxidation occurs in alkaline solution. **E**

The oxidation by air of cobalt(II) ions in ammoniacal solution is the experiment which led Alfred Werner to unravel the mysteries of what are now called *oxidation state* and *co-ordination number*.

The $[Co(NH_3)_6]^{3+}$ ion is yellow but the mixture is dark brown because of the presence of other species such as the purple ions $[Co(NH_3)_5(H_2O)]^{3+}$ and $[Co(NH_3)_5X]^{2+}$. X is the anion present in the original cobalt(II) salt.

hexaammine is rapidly oxidised to a dark brown mixture containing the hexaamminecobalt(III) ion:

$$[Co(H_2O)_6]^{2+} \xrightarrow{\text{NH}_3} [Co(NH_3)_6]^{2+} \xrightarrow{\text{O}_2} [Co(NH_3)_6]^{3+}$$

| pink | pale brown | yellow |
| octahedral | octahedral | octahedral |

14.4.6 *Catalysis*

A **catalyst** (see *Module 2, Foundation Physical and Inorganic Chemistry*, section 11.2.3) increases the rate of attainment of equilibrium, i.e. it makes reactions go faster. Catalysts are classified according to whether they act in a different phase from or in the same phase as the reactants.

E The phase may be solid or liquid or gas.

D A catalyst which acts in a **different phase** from the reactants is called a **heterogeneous** catalyst.

A catalyst which acts in the **same phase** as the reactants is called a **homogeneous** catalyst.

Because the success of an industrial process depends upon making the maximum amount of product in the shortest possible time, catalysis is at the heart of the chemical manufacturing industry. Most of the catalysts used in industry are transition metals or their compounds.

Heterogeneous catalysis
Heterogeneous catalysis depends on at least one reactant being adsorbed onto the catalyst surface (usually by weak chemical bonds), and being modified into a state which makes reaction more likely. Positions on the catalyst surface on which all such adsorption occurs are called **active sites**. If active sites are spaced appropriately, reactants can be held close enough to each other and in the right geometry to induce reaction.

Sometimes adsorption onto the surface can weaken reactant bonds, or induce a reactant molecule to break up into very reactive fragments; but equally, adsorption may be responsible for holding a reactant molecule on the surface in exactly the right configuration to make reaction easier.

In simple gas reactions such as

$A(g) + B(g) \rightarrow$ products

catalysis leading to an increase in reaction rate can come about if:

- **A** is adsorbed on the surface, becomes more accessible to collisions with **B**, and so reacts more readily than by random collisions in the gas phase. *Collision frequency increases.*

- **A** is held on the surface of the catalyst in a particularly favourable and highly reactive configuration. Unadsorbed **B** collides with reactive **A**. *Activation energy decreases.*

- **A** is adsorbed onto the surface and undergoes internal bond-breaking or rearrangement. Reactive fragments can easily react. *Activation energy decreases.*

Catalyst efficiency – strength of adsorption

Weak adsorption does not encourage reactants to come together; reaction rates stay low. _Strong adsorption_ keeps molecules immobile and fails to regenerate active sites; reaction rates fall if product molecules are not desorbed from the catalyst surface. The catalyst surface can become **poisoned** by product, rendering it ineffective.

Weak adsorption: Silver does not usually make an good catalyst because it fails to adsorb reactant molecules strongly enough, so that they never enjoy an enhanced opportunity to react together. Sometimes, weak adsorption can be used to advantage, so that silver can, exceptionally, be used as a _specific_ and _selective_ catalyst.

Strong adsorption: Tungsten is not a good catalyst because it adsorbs far too strongly and does not release products fast enough to increase the overall reaction rate. Tungsten has no commercial use as a catalyst.

Balanced adsorption: A good catalyst achieves a balance between efficient adsorption of reactants and effective desorption of products. Nickel and platinum achieve this balance, being both _specific_ and _selective_, and are therefore very versatile catalysts.

Making catalysts more effective

Heterogeneous catalysts act through surface properties. Increasing the surface area available for reaction is always worthwhile. A wafer-thin metal foil is much more active catalytically than is the same mass of metal in a solid lump. Expensive catalysts are frequently spread very thinly or impregnated onto an **inert support medium** in order to maximise the **surface-to-mass** ratio. Solids which have large surface areas in their own right (e.g. silica, asbestos) are useful as support media, especially if ground into very fine powders to maximise the surface area.

The practice of increasing surface area leads to more effective catalysis and also to the much more effective use of high-cost catalytic agents. **Loss of catalyst** from a support can greatly increase the inherent cost of a catalytic process, not least by decreasing the overall efficiency of the reaction.

Some catalysts can combine surface and support functions. The support itself may show catalytic activity and such **mixed-catalysts** can often perform dual functions.

The most useful catalysts, such as Pt and Rh, are also among the most expensive. **Catalytic converters** in cars (see _Module 3, Introduction to Organic Chemistry_, section 12.2.3) are designed to give a _high surface-area_ of active catalyst and use these metals spread thinly on a cheap ceramic support. Modern catalysts in car exhaust converters reduce noxious emissions to a tolerable level.

Weak catalytic activity can be beneficial, as in the oxidation of ethene to epoxyethane (see _Module 3, Introduction to Organic Chemistry_, section 12.3.3) where silver is used as a _weak_ but _effective_ catalyst, increasing the rate of one reaction without favouring other, unwanted, processes.

Metals at the beginning of a transition series (e.g. W) adsorb _too strongly_.

Metals near the end of a transition series (e.g. Ag) adsorb _too weakly_.

Metals in the middle of a transition series (e.g. Ni, Pd, Pt) balance these features and are _excellent catalysts_.

The Haber and Contact processes

Two very important industrial processes use heterogeneous catalysis by a transition metal or one of its compounds. In the production of ammonia from nitrogen and hydrogen by the **Haber process**, iron is the catalyst. In the manufacture of sulphuric acid by the **Contact process**, the conversion of sulphur dioxide into sulphur trioxide is catalysed by vanadium(V) oxide. The sulphur trioxide formed is dissolved in water to give sulphuric acid. These two processes are summarised below.

Haber process : $$N_2(g) + 3H_2(g) \underset{\overset{Fe(s)}{\rightleftharpoons}}{} 2NH_3(g)$$

Contact process : $$2SO_2(g) + O_2(g) \underset{\overset{V_2O_5(s)}{\rightleftharpoons}}{} 2SO_3(g)$$

The importance of the variable oxidation state of vanadium in the Contact process can be seen by considering how the catalyst works. Sulphur dioxide is oxidised to sulphur trioxide by vanadium(V) oxide:

$$SO_2 + V_2O_5 \rightarrow SO_3 + V_2O_4$$

The vanadium(V) has thus been reduced to vanadium(IV), which is then oxidised back to vanadium(V) oxide by oxygen:

$$2V_2O_4 + O_2 \rightarrow 2V_2O_5$$

Thus, the vanadium(V) oxide is unchanged at the end of the reaction and acts as a catalyst by use of the two oxidation states of vanadium, +5 and +4. The presence of the vanadium(V) oxide enables the reaction to proceed by a different route with a lower activation energy.

Catalyst poisons

Surface active catalysts are especially prone to **poisoning**. For example, catalytic converters for car exhausts are readily poisoned by lead *anti-knock* additives.

Poisoning occurs when unwanted contaminants, or waste products, become adsorbed too strongly onto the surface. Transition-metal catalysts, for example, are especially prone to poisoning by sulphur compounds which form inactive surface sulphides. The effect is cumulative. The sulphur content of feedstock gases in metal-catalysed reactions must be reduced to a minimum.

Homogeneous catalysis

Reactions which are catalysed homogeneously by a transition metal ion usually involve a change in oxidation state of the transition metal during catalysis. Such reactions proceed through the formation of an intermediate species which can form because the metal has variable oxidation states.

The oxidation of iodide ions by peroxodisulphate(VI) ions catalysed by iron ions

The reaction

$$S_2O_8^{2-} + 2I^- \rightarrow I_2 + 2SO_4^{2-}$$

is very slow even though it is energetically very favourable. Both ions are negatively charged and so are unlikely to make effective collisions with each other. The reaction occurs rapidly, however, when iron(II) ions are added. These cationic species can make effective collisions with anions.

Anti-knock additives such as tetraethyl-lead are no longer used in the UK.

E Catalyst poisons need to be removed from feedstock gases in the **Haber process**. The iron catalyst is poisoned by CO, CO_2 and water vapour, and also by sulphur compounds which are contaminants of the natural gas used to produce hydrogen.

Iron(II) ions are oxidised by peroxodisulphate(VI) ions; the iron(II) becomes iron(III) which, in turn, can rapidly oxidise iodide ions to iodine:

$$2Fe^{2+} + S_2O_8^{2-} \rightarrow 2Fe^{3+} + 2SO_4^{2-}$$
$$\underline{2Fe^{3+} + 2I^- \rightarrow 2Fe^{2+} + I_2}$$
$$S_2O_8^{2-} + 2I^- \rightarrow I_2 + 2SO_4^{2-}$$

The variable oxidation state of iron enables the reaction to proceed by an alternative route with a lower activation energy.

> **E**
> The two steps of the catalysed reaction can occur in either order, so that iron(III) ions can also be used as the catalyst.

Autocatalysis by manganese ions in the reaction between ethanedioate ions and manganate(VII) ions

When a solution of potassium manganate(VII), from a burette, is run into a hot, acidified solution containing ethanedioate ions, the purple colour is not decolorised immediately in the early stages of the titration. Once the initial purple colour produced by the addition of the first few drops of manganate(VII) solution has been discharged, however, further addition of the oxidant leads to immediate decolorisation. This rapid decolorisation continues until the end-point is reached.

The overall reaction is:

$$2MnO_4^- + 16H^+ + 5C_2O_4^{2-} \rightarrow 2Mn^{2+} + 10CO_2 + 8H_2O$$

The reaction between MnO_4^- ions and $C_2O_4^{2-}$ ions is slow; as in the previous example, both ions are negatively charged and so are less likely to make effective collisions. However, the reaction is catalysed by Mn^{2+} ions. Once some Mn^{2+} ions have been formed in the early stages of the titration, they can react readily with MnO_4^- ions to form Mn^{3+} ions:

$$4Mn^{2+} + MnO_4^- + 8H^+ \rightarrow 5Mn^{3+} + 4H_2O$$

These Mn^{3+} ions can then react with $C_2O_4^{2-}$ ions to liberate CO_2 and re-form Mn^{2+} ions:

$$2Mn^{3+} + C_2O_4^{2-} \rightarrow 2CO_2 + 2Mn^{2+}$$

So, until some Mn^{2+} ions have been formed, the manganate(VII) solution is decolorised only slowly. Such catalysis by a product of the same reaction is called **autocatalysis**.

14.4.7 *Other applications of transition-metal complexes*

The previous section has shown that it is the availability of variable oxidation states which makes an effective catalyst; this is why *transition metals and their compounds* find such widespread use in catalysis.

Transition-metal complexes in the body

As well as uses in catalysis, in industrial processes, and in chemical analysis, transition-metal complexes also play a vital, if less obvious, role in human biology.

Our bodies contain an iron(II) complex called haemoglobin; this complex is responsible for the red colour of blood and for the transport of oxygen by red blood cells from one part of the body to another. The iron(II) ion in

E See the picture of haem in section 14.4.2.

haemoglobin is octahedrally co-ordinated, with the porphyrin ligand providing four nitrogen donor atoms in a square plane around the iron; globin occupies a fifth position and the sixth position is occupied either by water or by molecular oxygen. The affinity of this complex for carbon monoxide is much greater than its affinity for oxygen, and it is the formation of very stable carboxyhaemoglobin, inhibiting the uptake of oxygen, that makes carbon monoxide so dangerous.

In medicine too, transition-metal complexes play important roles in drug formulation and use. Perhaps one of the best known examples is the anti-cancer drug **cisplatin**. This drug has had a remarkable success rate in the cure of certain types of cancer; its medicinal use is in chemotherapy. Cisplatin, whose structure is shown in Fig 15, is a four-coordinate, square-planar complex of platinum(II), with Cl^- ions and NH_3 molecules as ligands.

Fig 15
The structure of cisplatin

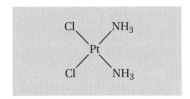

Cisplatin works by forming bonds with guanine bases in DNA, which causes cross-linking between DNA strands. Unwinding of the strands and replication is prevented, so that cancerous cells cannot divide and grow.

Uses of silver complexes

In section 14.4.3, it was noted that silver(I) ions usually form linear complexes. The uses of each of the complexes mentioned previously are considered below.

E See Module 3, Introduction to Organic Chemistry, section 12.5.2.

$[Ag(NH_3)_2]^+$ This complex ion is formed when silver chloride or silver oxide dissolves in aqueous ammonia to form the colourless solution known as Tollens' reagent. This reagent is used to distinguish aldehydes from ketones; aldehydes give a silver mirror on the side of the test tube, ketones do not:

$$RCHO + 2[Ag(NH_3)_2]^+ + 3OH^- \rightarrow RCOO^- + 2Ag + 4NH_3 + 2H_2O$$

$[Ag(S_2O_3)_2]^{3-}$ This complex ion is formed when silver salts are dissolved in sodium thiosulphate ('hypo') solution. The reaction is important in photographic fixing. Silver bromide which has not been exposed to light is dissolved away from the film, leaving a black image of silver as the negative:

$$AgBr + 2S_2O_3^{2-} \rightarrow [Ag(S_2O_3)_2]^{3-} + Br^-$$

The deposition potential for silver from the cyano-complex is greatly raised over that for aqueous silver solutions. Thus, silver can only be displaced if a sufficiently high potential is applied. Under such conditions, the silver deposits in a form which sticks firmly to the base metal being plated, hence its use in silver-plating.

$[Ag(CN)_2]^-$ This complex ion is formed when silver salts are dissolved in aqueous sodium or potassium cyanide. The resulting solution is used as the electrolyte in silver-plating.

14.5 Reactions of inorganic compounds in aqueous solution

14.5.1 *Lewis acids and bases*

In 1938, G.N. Lewis proposed a definition of acids and bases which has become an important concept in chemistry:

> **A Lewis acid is *an electron-pair acceptor***
>
> **A Lewis base is *an electron-pair donor***
>
> **D**

Molecules or ions with lone pairs (i.e. non-bonded pairs) of electrons are capable of behaving as bases, and molecules or ions which can accept a pair of electrons are capable of behaving as acids. For example, in the reaction between H_2O and H^+:

> H_2O has two *lone pairs* of electrons on the oxygen atom, one of which it can *donate*.
>
> H_2O can act as a *Lewis base*.

> H^+ has *no electrons*, so it can *accept* an electron pair (to achieve a full shell).
>
> H^+ can act as a *Lewis acid*.

Consequently, the reaction which occurs between H_2O and H^+ is:

$$H_2O + H^+ \rightarrow H_3O^+$$

Another example is the reaction of boron trifluoride with ammonia:

$$BF_3 + NH_3 \rightarrow [F_3B{\leftarrow}NH_3]$$

Example 26

Try to work out which is the acid and which the base in the reaction between BF₃ and NH₃.

Method

Inspect the equation

$$BF_3 + NH_3 \rightarrow [F_3B{\leftarrow}NH_3]$$

to find which species is acting as a *lone-pair donor*. This species is the *Lewis base*.

Answer

Ammonia has a lone pair of electrons on the nitrogen atom, and so is acting as a base. Boron in boron trifluoride has only six electrons in its valence shell and is capable of accepting two more electrons to give a full shell of eight; it can therefore act as an acid.

Comment

When a Lewis acid reacts with a Lewis base, a co-ordinate bond is formed. A straight arrow is drawn to show the origin of the electron pair in the newly formed co-ordinate bond.

Because it is clear that one electron pair on nitrogen has been donated to boron, it can be helpful to show this process by means of a straight arrow.

In the case of H_3O^+, however, the oxygen atom shares three electron pairs with three *equivalent* hydrogen atoms, so an arrow cannot be drawn to single out any one of them; all three O–H bonds are identical, unlike the B–N bond, which is clearly the only *donor–acceptor* bond in the molecule.

Note the similarity of the straight arrow to the *curly arrow* used in organic chemistry (see *Module 3, Introduction to Organic Chemistry*, section 12.3.2). The straight arrow also starts at a pair of electrons and points to where they end up in the new species.   **E**

Remember that:
Lewis base = nucleophile
= ligand

In transition metal chemistry, all **ligands** are *Lewis bases* and all metal cations (having a positive charge) are *Lewis acids*. The ligand *donates a pair of electrons* to the metal.

14.5.2 *Metal-aqua ions*

When a transition-metal ion is placed in water, one of the two lone pairs of electrons on each of six water molecules is used to form a co-ordinate bond with the metal ion. The result is a **co-ordination compound** or **complex**. This reaction and its associated colour change was illustrated in section 14.4.2, using copper(II) sulphate.

A still more striking colour change occurs when anhydrous cobalt(II) chloride is added to water. The anhydrous salt is blue whereas the hexaaquacobalt(II) cation is pink.

The colours of solutions of transition metal salts in water are largely due to the colours of the metal-aqua ions, e.g.

$[Fe(H_2O)_6]^{2+}$ green

$[Co(H_2O)_6]^{2+}$ pink

$[Cu(H_2O)_6]^{2+}$ blue

Any copper(II) salt in aqueous solution will be present as the hexaaqua ion and the solution will be blue. Anhydrous copper(II) salts are not blue: $CuSO_4$ is white, $CuCl_2$ is yellow, $CuBr_2$ is black, and CuO is also black, yet they all give blue solutions when dissolved in water or dissolved in dilute acids.

So strong is the bonding between the water molecules and the metal ions, that the hexaaqua ion remains intact when solutions containing this ion are evaporated. Thus, solid transition-metal salts in the laboratory are often not simple salts but contain a complex, the metal-hexaaqua ion. Examples of 'salts' containing the hexaaqua ion in the solid state include:

$FeSO_4.7H_2O$	green	contains $[Fe(H_2O)_6]^{2+}$
$CoCl_2.6H_2O$	pink	contains $[Co(H_2O)_6]^{2+}$
$Fe(NO_3)_3.9H_2O$	pale violet	contains $[Fe(H_2O)_6]^{3+}$

Metal(III)-aqua ions

Whilst the reactions of metal(II) salts with water do not appear to be violent, enough energy is given out on hydration to overcome the lattice energy of the solid so that the solid dissolves; reaction with water breaks down the crystal lattice. When the charge on the metal ion is increased to 3+, however, the hydration reaction is often very exothermic. Thus, aluminium chloride dissolves in water violently with evolution of heat and fumes:

$$AlCl_3(s) \xrightarrow{\text{excess } H_2O(l)} [Al(H_2O)_6]^{3+}(aq) + 3Cl^-(aq)$$

Note that there is no precipitate of a metal hydroxide in this reaction; in general, when simply added to water, metal(III) chlorides do not hydrolyse to form metal(III) hydroxides.

E Note that the anion plays no part in the final colour seen. If green cobalt(II) bromide is added to water, the same pink hexaaquacobalt(II) cation is formed as with cobalt(II) chloride. The bromide and chloride ions in these reactions become aquated, forming colourless $Br^-(aq)$ ions and $Cl^-(aq)$ ions, respectively.

Metal(IV) chlorides, however, do hydrolyse further and a precipitate of the hydroxide is often seen, e.g. $TiCl_4$ gives $Ti(OH)_4$.

Metal ions exist in aqueous solution as 6-water or hexaaqua ions. The hexaaqua ion contains a metal ion surrounded octahedrally by six oxygen atoms from six water molecules. When metal ions react in aqueous solution, it is essential to remember these six water molecules because they can take part in the chemical reactions of metal ions.

When chemical reactions occur, bonds are frequently broken and, more often than not, new bonds are formed. The driving force for reaction is often bond formation. When considering reactions in water, the questions to ask are 'which bonds can be broken' and 'which bonds can be formed'?

The bonding in a hexaaqua ion, $[M(H_2O)_6]^{n+}$ is shown in Fig 16.

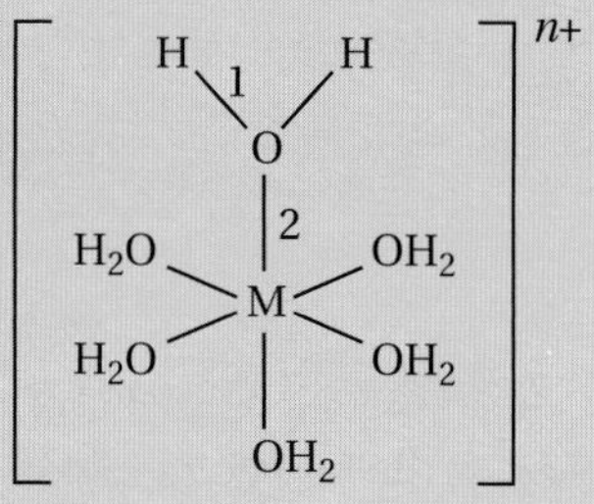

Fig 16
The hexaaqua ion $[M(H_2O)_6]^{n+}$ contains only *two* types of bond:

O–H bonds, covalent bonds in water molecules co-ordinated to the metal ion; one of these bonds is labelled 1.

M–O bonds, co-ordinate bonds between water ligands and the metal ion; one of these bonds is labelled 2.

In reactions of metal ions, the only process which occurs without breaking either of these two types of bond is a redox reaction, i.e. oxidation or reduction, which involves only the loss or gain of an electron, e.g.

$$[M(H_2O)_6]^{2+} \rightarrow [M(H_2O)_6]^{3+} + e^-$$

If the O–H bond in a co-ordinated water molecule is broken, the reaction is called an **acidity reaction** or a **hydrolysis reaction**; if the M–O bond is broken (and a new ligand is attached to the metal) the reaction is called a **substitution reaction**. Both of these important reactions are considered in more detail below.

14.5.3 *The acidity or hydrolysis reaction*

When a metal-aqua ion is placed in water, an equilibrium is set up

$$[M(H_2O)_6]^{2+} + H_2O \rightleftharpoons [M(H_2O)_5(OH)]^+ + H_3O^+$$

which is called the *hydrolysis* reaction because a water molecule has been split into OH^- and H^+. It is also called the *acidity* reaction because hydrolysis leads to the formation of H_3O^+ ions.

For a metal(II) ion, as above, the equilibrium lies very much to the left-hand side, so that metal(II) ions are only slightly acidic in water. The degree of acidity is measured by the magnitude of the equilibrium constant for the reaction. In *Module 4, Further Physical and Organic*

Chemistry, section 13.3.5, the *acid dissociation constant* (or acidity constant) is defined as

$$K_a = \frac{[M(H_2O)_5(OH)^+][H_3O^+]}{[M(H_2O)_6^{2+}]}$$

E $pK_a = -\log_{10}K_a$

For metal(II) ions, K_a varies between 10^{-6} and 10^{-11} and pK_a varies between 6 and 11.

For metal(III) ions, however, the equilibrium

$$[M(H_2O)_6]^{3+} + H_2O \rightleftharpoons [M(H_2O)_5(OH)]^{2+} + H_3O^+$$

lies further to the right and, for this equilibrium, K_a varies between 10^{-2} and 10^{-5}, so that pK_a takes values between 2 and 5.

For metal(IV) ions, the hydrolysis is so great that this first stage of the equilibrium goes virtually to completion ($pK_a < 1$).

The acidity of transition-metal ions at usual concentrations can be summarised as follows:

$[M(H_2O)_6]^{2+}$	$[M(H_2O)_6]^{3+}$
very weak acids	weak acids
pH around 6	pH around 3

The principal species present in aqueous solutions of metal(II) and metal(III) salts are, therefore, the hexaaqua ions. For metal(II) ions, only about one in ten thousand aqua ions has undergone the hydrolysis reaction whereas for metal(III) ions, more than one in a thousand has been hydrolysed.

Because of the extreme hydrolysis of metal(IV) ions, the ion $[M(H_2O)_6]^{4+}$ does not usually exist in aqueous solution; it hydrolyses to $M(OH)_4$.

Two factors are involved in deciding the acidity of metal ions; these are:

- the charge on the metal ion   *acidity increases with charge*
- the size of the metal ion   *acidity decreases as size increases*

E Solutions containing $[M(H_2O)_6]^{3+}$ ions are more acidic than those containing $[M(H_2O)_6]^{2+}$ ions because M^{3+} ions are more polarising than M^{2+} ions.

Small, highly-charged cations are the strongest acids in aqueous solution. If the charge and size factors are taken together, the **charge to size ratio** can be used to predict the relative acidities of metal ions. This ratio reflects the polarising power of the metal ion. The greater the power of the metal ion to attract electron density from the oxygen atom of a co-ordinated water molecule, the weaker is the O–H bond in this molecule. As a result, it is easier to break this O–H bond and release an H^+ ion to a water molecule which is not co-ordinated.

So far, only the first stage of the hydrolysis reaction has been considered. Further hydrolysis can occur (as is seen with metal(IV) ions) and this is especially true if a base is added to upset the equilibrium.

For metal(II) ions the complete hydrolysis is shown by the equations

$$[M(H_2O)_6]^{2+} + H_2O \rightleftharpoons [M(H_2O)_5(OH)]^+ + H_3O^+$$
$$[M(H_2O)_5(OH)]^+ + H_2O \rightleftharpoons [M(H_2O)_4(OH)_2] + H_3O^+$$

E Bases react with transition metal ions to form precipitates of metal hydroxides; further reactions may occur if an excess of base is added (see *Amphoteric character* below).

while for metal(III) ions, the equations are:

$$[M(H_2O)_6]^{3+} + H_2O \rightleftharpoons [M(H_2O)_5(OH)]^{2+} + H_3O^+$$

$$[M(H_2O)_5(OH)]^{2+} + H_2O \rightleftharpoons [M(H_2O)_4(OH)_2]^+ + H_3O^+$$

$$[M(H_2O)_4(OH)_2]^+ + H_2O \rightleftharpoons [M(H_2O)_3(OH)_3] + H_3O^+$$

Note that the final product in all cases is the neutral metal hydroxide. This species has no overall charge and there is therefore little hydration energy available to overcome the lattice energy, so that such hydroxides always appear as precipitates.

- All transition metal hydroxides are insoluble in water; they are only formed from aquated metal(II) and metal(III) ions when a base (e.g. NaOH or NH_3) is added.

Reactions of aqua ions with water

Some reactions, which can only be understood by using these hydrolysis equations, will now be considered. If either some pale violet crystals of iron(III) nitrate nonahydrate, $Fe(NO_3)_3.9H_2O$, or some crystals of iron(III) ammonium alum, $Fe_2(SO_4)_3.(NH_4)_2SO_4.24H_2O$, are added to water, a brown solution forms. Both of these iron salts contain the very pale violet hexaaqua ion $[Fe(H_2O)_6]^{3+}$, and it is clearly this ion which is reacting with the water:

$$[Fe(H_2O)_6]^{3+} + H_2O \rightleftharpoons [Fe(H_2O)_5(OH)]^{2+} + H_3O^+$$

pale violet   brown

The hydrolysis product is brown and, since this ion is more intensely coloured than the hexaaqua ion, the colour of the solution is seen to be brown. Don't forget, however, that the above equilibrium still lies very far to the left, so that $[Fe(H_2O)_6]^{3+}$ remains the principal species present.

Applying the principles of equilibrium (see *Module 2, Foundation Physical and Inorganic Chemistry*, section 11.3.2) to this case, allows prediction of the outcome when either acid or base is added to the system above.

If a little nitric acid is added to the brown solution, the brown colour disappears and an almost colourless (pale violet) solution results. Clearly, if an acid is added to the equilibrium system above, the equilibrium will respond by moving to the left to remove the excess of acid, and some $[Fe(H_2O)_5(OH)]^{2+}$ will be converted into $[Fe(H_2O)_6]^{3+}$. Similarly, if a base such as sodium hydroxide is added to the equilibrium system above, the base will react with H_3O^+ ions, removing some of them; the system will respond by producing more H_3O^+ ions and the equilibrium will move to the right. What is observed on adding acid, therefore, is a decrease of the dark brown colour. On adding base, the solution turns darker brown until eventually (when the next two stages of the hydrolytic equilibria have been moved to the right) a brown precipitate of iron(III) hydroxide forms.

Reactions of aqua ions with alkalis

When sodium hydroxide solution is added to a solution of a transition metal salt, a precipitate of the metal hydroxide always appears.

> **E**
>
> Metal hydroxides can be written as $M(OH)_2$ and $M(OH)_3$. Whilst the hydroxides $M(OH)_2$ do actually occur with this precise formula in the solid state, metal(III) hydroxides readily lose water on drying and usually occur in the solid state as MO(OH). The compound with the formula FeO(OH), for example, is 'rust'.

For instance, consider what happens when sodium hydroxide solution is added to a solution of copper(II) sulphate. The species present in a solution of copper(II) sulphate are:

$[Cu(H_2O)_6]^{2+}$ (major species) $[Cu(H_2O)_5(OH)]^+$ and H_3O^+ (small amount)

When OH^- ions are added to this mixture, they attack the strongest acid present, which is H_3O^+:

$$OH^- + H_3O^+ \rightarrow 2H_2O$$

This reaction upsets the hydrolysis equilibrium

$$[Cu(H_2O)_6]^{2+} + H_2O \rightleftharpoons [Cu(H_2O)_5(OH)]^+ + H_3O^+$$

and, when this equilibrium has moved completely over to the right-hand side, the next equilibrium is set up:

$$[Cu(H_2O)_5(OH)]^+ + H_2O \rightleftharpoons [Cu(H_2O)_4(OH)_2] + H_3O^+$$

As more sodium hydroxide is added, further H_3O^+ is removed and blue copper(II) hydroxide precipitates. Thus, the overall equation is:

$$[Cu(H_2O)_6]^{2+} + 2OH^- \rightleftharpoons [Cu(OH)_2(H_2O)_4] + 2H_2O$$

The same sequence occurs whatever the metal ion, so that the reactions of sodium hydroxide with transition metal ions can be summarised as:

- The addition of *sodium hydroxide* to a solution of a transition metal salt *always precipitates* the transition metal hydroxide.

The colours of some common hydroxides formed in these reactions are:

$[Fe(H_2O)_6]^{2+}$	$\rightarrow$	$[Fe(H_2O)_4(OH)_2]$
green solution		green precipitate
$[Co(H_2O)_6]^{2+}$	$\rightarrow$	$[Co(H_2O)_4(OH)_2]$
pink solution		blue-green precipitate
$[Fe(H_2O)_6]^{3+}$	$\rightarrow$	$[Fe(H_2O)_3(OH)_3]$
pale violet solution		brown precipitate
$[Cr(H_2O)_6]^{3+}$	$\rightarrow$	$[Cr(H_2O)_3(OH)_3]$
red-violet solution		green precipitate

Reactions of aqua ions with aqueous ammonia

An aqueous solution of ammonia is alkaline by virtue of the equilibrium:

$$NH_3 + H_2O \rightleftharpoons NH_4^+ + OH^-$$

Thus, because of the presence of OH^- ions in the ammonia solution, the addition of aqueous ammonia to a solution of a transition metal salt results in exactly the same initial sequence of events as occur on the addition of sodium hydroxide:

- The addition of ammonia solution to a solution of a transition metal salt *always* results, *initially*, in the formation of a precipitate of the metal hydroxide.

A further reaction (i.e. *substitution*, see section 14.5.4) can occur in the presence of an excess of ammonia, but initially a *precipitate* of the metal hydroxide is always formed.

E If insufficient base is added, the hydrolysis will only proceed as far as the formation of $[Cu(H_2O)_5OH)]^+$. This cation can precipitate with an anion as a basic salt, e.g. $[Cu(H_2O)_5(OH)]_2SO_4$, or $CuSO_4.Cu(OH)_2.10H_2O$, which is seen as a pale blue precipitate.

E Iron(II) hydroxide is actually white but darkens rapidly in air due to oxidation.

E Cobalt(II) hydroxide exists in two forms; the blue-green form turns slowly into the more stable pink form at room temperature.

Reactions of aqua ions with carbonate ions

Sodium carbonate is commonly used as a base, so it is important to be aware of its reactions with metal ions. The carbonate ion reacts with acids to form the hydrogencarbonate ion and then carbonic acid (H_2CO_3), which rapidly becomes carbon dioxide and water:

$$CO_3^{2-} + H_3O^+ \rightleftharpoons HCO_3^- + H_2O$$
$$HCO_3^- + H_3O^+ \rightleftharpoons CO_2 + 2H_2O$$

Metal(II) ions are not sufficiently acidic to displace carbonic acid from carbonates. Instead, these ions form insoluble metal carbonates, $MCO_3(s)$.

- The addition of sodium carbonate solution to a solution of a metal(II) salt results in a *precipitate* of the metal(II) carbonate.

Metal(III) ions, however, are more acidic than carbonic acid and therefore displace this acid from solutions containing carbonate ions (a strong acid will always displace a weaker one). So the reaction between a solution of metal(III) ions and carbonate ions results in *effervescence* and *precipitation* of the metal hydroxide.

The reaction follows the scheme

$$[M(H_2O)_6]^{3+} + H_2O \rightleftharpoons [M(H_2O)_5(OH)]^{2+} + H_3O^+$$

followed by the reaction of H_3O^+ with CO_3^{2-}

$$2H_3O^+ + CO_3^{2-} \rightarrow CO_2 + 3H_2O$$

so that the equilibrium is pushed to the right and subsequent equilibria are set up

$$[M(H_2O)_5(OH)]^{2+} + H_2O \rightleftharpoons [M(H_2O)_4(OH)_2]^+ + H_3O^+$$
$$[M(H_2O)_4(OH)_2]^+ + H_2O \rightleftharpoons [M(H_2O)_3(OH)_3] + H_3O^+$$

until the metal(III) hydroxide precipitates.

In general:

- The reaction between a metal(III) salt solution and sodium carbonate leads to *evolution of a gas* (carbon dioxide) and *precipitation* of the metal(III) hydroxide.

As a result, metal(III) carbonates cannot be prepared in aqueous solution and, in general, they do not exist. So, for example, $Al_2(CO_3)_3$, $Fe_2(CO_3)_3$ and $Cr_2(CO_3)_3$ *cannot* be prepared in aqueous solution and are, indeed, unknown.

The three equations in the hydrolysis of a metal(III) ion can be added together to give the overall equation:

$$[M(H_2O)_6]^{3+} + 3H_2O \rightleftharpoons [M(H_2O)_3(OH)_3] + 3H_3O^+$$

This equilibrium is driven to the right following removal of H_3O^+ by carbonate ions:

$$2H_3O^+ + CO_3^{2-} \rightarrow CO_2 + 3H_2O$$

Combining these two equations gives:

$$2[M(H_2O)_6]^{3+} + 3CO_3^{2-} \rightarrow 2[M(H_2O)_3(OH)_3] + 3CO_2 + 3H_2O$$

It is not only carbonate ions which behave in this way. Anions derived from weak acids are protonated by metal(III) aqua ions so that the free acid and the metal(III) hydroxide result.

Some examples of anions of weak acids are S^{2-}, NO_2^-, SO_3^{2-}, $S_2O_3^{2-}$ and CH_3COO^-. Metal(III) salts of these anions cannot usually be prepared in aqueous solution.

Amphoteric character

All of the hydrolytic equilibria given above can be reversed by using strong acids. Thus, metal hydroxides dissolve in strong acids to give metal-aqua ions (as long as a non-complexing strong acid such as HNO_3 is used; nitrate ions are only *weak* ligands).

Whilst metal hydroxides do not undergo hydrolysis in water, they can be attacked by the strong nucleophile OH^- to give anionic complexes which are water-soluble. This ability of metal hydroxides to dissolve in both strong acids and strong alkalis is known as **amphoteric character**.

When sodium hydroxide is added to a solution of an aluminium salt, a white precipitate of aluminium hydroxide is seen initially. Then, as an excess of the alkali is added, the precipitate dissolves to form a colourless solution containing the aluminate ion:

$$[Al(H_2O)_6]^{3+} + 3H_2O \rightleftharpoons [Al(H_2O)_3(OH)_3] + 3H_3O^+$$

$$[Al(H_2O)_3(OH)_3] + OH^- \rightleftharpoons [Al(OH)_4]^- + 3H_2O$$

Similarly, the addition of sodium hydroxide to a solution of a chromium(III) salt leads to the initial formation of a green precipitate of chromium(III) hydroxide which then dissolves to give a green solution containing chromate(III) ions when an excess of the alkali is added:

$$[Cr(H_2O)_6]^{3+} + 3H_2O \rightleftharpoons [Cr(H_2O)_3(OH)_3] + 3H_3O^+$$

$$[Cr(H_2O)_3(OH)_3] + 3OH^- \rightleftharpoons [Cr(OH)_6]^{3-} + 3H_2O$$

The general scheme for metal(III) ions is:

$$[M(H_2O)_6]^{3+} \underset{H_3O^+}{\overset{OH^-}{\rightleftharpoons}} [M(H_2O)_3(OH)_3] \underset{H_3O^+}{\overset{OH^-}{\rightleftharpoons}} [M(OH)_4]^-$$

acidic solution   neutral solution   alkaline solution

Most transition-metal hydroxides are amphoteric.

In high oxidation states, transition metals often exist in solution in an anionic form. An important anionic equilibrium occurs in chromium(VI) chemistry. Chromate(VI) ions (see section 14.4.5) react with acids to form dichromate(VI) ions:

$$2CrO_4^{2-} + 2H^+ \rightleftharpoons Cr_2O_7^{2-} + H_2O$$

yellow   orange
chromate(VI)   dichromate(VI)

The dichromate(VI) ion is converted back into the chromate(VI) ion in alkaline solution:

$$Cr_2O_7^{2-} + 2OH^- \rightleftharpoons 2CrO_4^{2-} + H_2O$$

Summaries of the reactions of metal(II) and metal(III) ions with bases are given in Tables 19 and 20 on page 67.

E The ending *-ate* means that the metal is present in a negatively-charged ion (an anion).

E Whereas *all bases* yield hydroxide precipitates, only *strong bases in excess* can cause some $M(H_2O)_3(OH)_3$ precipitates to re-dissolve, forming negatively-charged complexes. An excess of a *weak base* does not re-dissolve these hydroxide precipitates.

E Chromium(III) is unusual in that it never forms tetrahedral complexes, so that the species $[Cr(OH)_4]^-$ does not exist.

Some hydroxides are polymeric and require a solution of sodium hydroxide stronger than 'dilute' in order to dissolve. Adding a solution of copper(II) sulphate to a concentrated solution of sodium hydroxide yields a deep-blue solution of cuprate(II) ions, with no trace of precipitate.

Both ions contain chromium surrounded tetrahedrally by oxygen: dichromate(VI) has an oxygen bridge between the two chromium atoms.

14.5.4 *Substitution reactions*

Each of the acidity reactions above has involved the breaking of an O–H bond in a co-ordinated water molecule. Other reactions in which one or more of the M–O bonds in the hexaaqua ion are broken will now be considered. When a water molecule is replaced by another ligand, a substitution reaction occurs; it is *nucleophilic substitution*, but is often just called **ligand substitution**.

For the purposes of substitution reactions, there are two types of ligand:

- *neutral* – uncharged molecules
- *anionic* – negatively-charged ligands

Substitution by neutral ligands

Ammonia can act as a base in the *Brønsted–Lowry* sense (when it reacts with a proton, as it does in the acidity reaction considered above) and also as a base in the *Lewis* sense (when it acts as a ligand).

A general equation can be written for the replacement of water molecules in an aqua ion by neutral ligands. For example, the replacement of H_2O by NH_3 can be written as:

$$[M(H_2O)_6]^{2+} + 6NH_3 \rightleftharpoons [M(NH_3)_6]^{2+} + 6H_2O$$

The equation above disguises the fact that what is written as a one-step equilibrium can be broken down into six steps, with only one water molecule being replaced in each step:

$$[M(H_2O)_6]^{2+} + NH_3 \rightleftharpoons [M(NH_3)(H_2O)_5]^{2+} + H_2O$$

$$[M(NH_3)(H_2O)_5]^{2+} + NH_3 \rightleftharpoons [M(NH_3)_2(H_2O)_4]^{2+} + H_2O$$

$$[M(NH_3)_2(H_2O)_4]^{2+} + NH_3 \rightleftharpoons [M(NH_3)_3(H_2O)_3]^{2+} + H_2O$$

$$[M(NH_3)_3(H_2O)_3]^{2+} + NH_3 \rightleftharpoons [M(NH_3)_4(H_2O)_2]^{2+} + H_2O$$

$$[M(NH_3)_4(H_2O)_2]^{2+} + NH_3 \rightleftharpoons [M(NH_3)_5(H_2O)]^{2+} + H_2O$$

$$[M(NH_3)_5(H_2O)]^{2+} + NH_3 \rightleftharpoons [M(NH_3)_6]^{2+} + H_2O$$

At first sight, these equations look complicated, but they are in fact quite simple. The first equation starts with a hexaaqua ion and the last equation ends with a hexaammine ion. Note that, when water is bonded to a metal ion, it is called *aqua*, but when ammonia is bonded, it is called *ammine*.

Ammonia is uncharged and has a similar size to water. Consequently, no change of shape is expected to occur during these substitution reactions. Thus, all the complexes in the equilibria above are *octahedral*.

One of the most famous historical examples of such a substitution sequence occurs in the addition of ammonia to cobalt(II) ions. In this sequence, the first change observed is the formation of a *green-blue precipitate*. This precipitate is cobalt(II) hydroxide, formed by the acidity reaction of the cobalt(II) ions. The overall reaction for this process can be written as:

$$[Co(H_2O)_6]^{2+} + 2NH_3 \rightleftharpoons [Co(H_2O)_4(OH)_2] + 2NH_4^+$$
pink solution $\qquad\qquad$ green-blue precipitate

> **E** Note the double 'm' in *ammine*; an *amine* is something quite different (see *Module 4, Further Physical and Organic Chemistry*, section 13.7).

> From a study of isomerism in the resulting cobaltammines in this reaction, Alfred Werner realised that metals could show two types of valency, nowadays called *oxidation state* and *co-ordination number*.

When an excess of concentrated aqueous ammonia is added to the reaction mixture, the green-blue precipitate dissolves and a *pale straw-coloured solution* results. It is important to keep air away from this solution, which otherwise darkens rapidly to form a *dark-brown mixture* containing cobalt(III) ammines (see section 14.4.5).

The species in the straw-coloured solution is the hexaamminecobalt(II) ion. The green-blue hydroxide dissolves in ammonia according to the equation:

$$[Co(H_2O)_4(OH)_2] + 6NH_3 \rightleftharpoons [Co(NH_3)_6]^{2+} + 4H_2O + 2OH^-$$
green-blue precipiate   straw-coloured solution

The overall equation starting from the aqua ion is:

$$[Co(H_2O)_6]^{2+} + 6NH_3 \rightleftharpoons [Co(NH_3)_6]^{2+} + 6H_2O$$
pink solution   straw-coloured solution

Thus, with cobalt(II), complete substitution of water molecules by ammonia molecules occurs.

With copper(II) ions, however, only four of the six water molecules on copper are replaced. As with cobalt, when aqueous ammonia is added to a solution of a copper(II) salt, the first change observed is the formation of a *blue precipitate* of the hydroxide. This then dissolves when an excess of ammonia is added and a *deep-blue solution* of the tetraamminebisaquacopper(II) ion is formed.

$$[Cu(H_2O)_6]^{2+} + 4NH_3 \rightleftharpoons [Cu(NH_3)_4(H_2O)_2]^{2+} + 4H_2O$$
blue solution   deep-blue solution

Further substitution can be achieved when the concentration of ammonia is increased (e.g. by cooling the solution in ice and saturating it with ammonia gas, or by using liquid ammonia rather than aqueous ammonia), but in concentrated aqueous ammonia, the equilibrium position reached is that in which the dark-blue $[Cu(NH_3)_4(H_2O)_2]^{2+}$ ion is formed. This ion has four ammonia molecules in a square-planar arrangement around copper with water molecules occupying the other two octahedral positions above and below the plane. The bonds to water are longer and weaker than the bonds to ammonia.

A summary of the reactions of metal(II) and metal(III) ions with various bases (hydroxide ions, ammonia and carbonate ions) is given in Tables 19 and 20 opposite.

A more sophisticated explanation for the lack of formation of the hexaammine, based on the Jahn–Teller effect, can be found in undergraduate texts.

Base added	Aqueous M(II) ions		
	$[Fe(H_2O)_6]^{2+}$ green solution	$[Co(H_2O)_6]^{2+}$ pink solution	$[Cu(H_2O)_6]^{2+}$ blue solution
OH⁻ (little)	$[Fe(H_2O)_4(OH)_2]$ green ppt (turns brown in air)	$[Co(H_2O)_4(OH)_2]$ blue ppt (turns pink on standing and then turns brown in air)	$[Cu(H_2O)_4(OH)_2]$ blue ppt
OH⁻ (excess dilute)			
NH₃ (little)			
NH₃ (excess)	as above	$[Co(NH_3)_6]^{2+}$ pale-brown (straw) solution (turns brown in air)	$[Cu(NH_3)_4(H_2O)_2]^{2+}$ deep-blue solution
CO_3^{2-}	$FeCO_3$ green ppt	$CoCO_3$ pink ppt	$CuCO_3$ green-blue ppt

Table 19
Summary of the reactions of metal(II) ions with bases

Base added	Aqueous M(III) ions		
	$[Fe(H_2O)_6]^{3+}$ violet solution (appears yellow due to hydrolysis)	$[Al(H_2O)_6]^{3+}$ colourless solution	$[Cr(H_2O)_6]^{3+}$ dull red-blue solution (ruby)
OH⁻ (little)	$[Fe(H_2O)_3(OH)_3]$ brown ppt	$[Al(H_2O)_3(OH)_3]$ white ppt	$[Cr(H_2O)_3(OH)_3]$ green ppt
OH⁻ (excess)	as above brown ppt	$[Al(OH)_4]^-$ colourless solution	$[Cr(OH)_6]^{3-}$ green solution
NH₃ (little)	as above brown ppt	$[Al(H_2O)_3(OH)_3]$ white ppt	$[Cr(H_2O)_3(OH)_3]$ green ppt
NH₃ (excess)	as above brown ppt	as above white ppt	$[Cr(NH_3)_6]^{3+}$ purple solution
CO_3^{2-}	as above brown ppt *and* effervescence of CO_2	as above white ppt *and* effervescence of CO_2	$[Cr(H_2O)_3(OH)_3]$ green ppt *and* effervescence of CO_2

Table 20
Summary of the reactions of metal(III) ions with bases

Substitution reactions involving Cr(III) complexes are very slow; the formation of $[Cr(NH_3)_6]^{3+}$ under laboratory conditions can take several minutes, whereas $[Co(NH_3)_6]^{2+}$ forms immediately.

Substitution by chloride ions

All anions are potentially capable of acting as ligands. For example, the very common ligand Cl⁻ has the electron arrangement of the noble gas argon; it has four lone pairs of electrons. When Cl⁻ bonds to a metal ion, one of these lone pairs forms a co-ordinate bond to the metal; the three remaining lone pairs on the chloride ion do not co-ordinate.

E Concentrated hydrochloric acid contains about 11 mol dm^{-3} of HCl and is virtually completely dissociated:
$$HCl + H_2O \rightarrow H_3O^+ + Cl^-$$

E Note that a high concentration of chloride ions is necessary to push the equilibrium over to the right-hand side; it is hard to achieve so high a concentration even with saturated aqueous sodium chloride.

A good source of chloride ions is concentrated hydrochloric acid. Hydrogen chloride is very much more soluble in water than ionic chlorides such as sodium chloride, allowing high concentrations to be achieved; this is why concentrated hydrochloric acid is used to prepare complexes of chloride ions with transition metals.

When a pink solution of a cobalt(II) salt in water is treated with an excess of concentrated hydrochloric acid, a deep-blue solution is formed:

$$[Co(H_2O)_6]^{2+} + 4Cl^- \rightleftharpoons [CoCl_4]^{2-} + 6H_2O$$
pink   blue

octahedral   tetrahedral

When the blue solution of $[CoCl_4]^{2-}$ is diluted with water, the pink colour returns. This behaviour can be understood in terms of an equilibrium which can be driven from left to right by high chloride ion concentrations and from right to left if the concentration of chloride ions is lowered by dilution. This marked change in colour is brought about not only by the change of ligand but, more significantly, also by the change in the co-ordination number of the cobalt ion. The shape of the ligands around the cobalt ion changes from octahedral to tetrahedral as hydrochloric acid is added.

Why then does the shape of the complex ion change? The chloride ligand is

- negatively charged
- and large

As the size of the ligands around a metal ion is increased, a point is reached when the electron charge-clouds around the ligands repel each other to such an extent that the octahedral structure becomes less stable than the tetrahedral one. In the tetrahedral structure, the ligands are approximately 109° apart and do not experience repulsive forces as great as in the octahedral arrangement, where the ligands are only 90° apart.

E If the size of the ligand is decreased to the smaller F$^-$ ion, then octahedral complexes become common, e.g. AlF_6^{3-} (c.f. $AlCl_4^-$). If, however, the size of the ligand is increased to the larger I$^-$ ion, octahedral complexes are not formed with any of the cations of the first three periods.

Thus, the general equation for the substitution reaction of a metal(II)-aqua ion with chloride ions is:

$$[M(H_2O)_6]^{2+} + 4Cl^- \rightleftharpoons [MCl_4]^{2-} + 6H_2O$$
octahedral   tetrahedral

Example 27

Work out what will happen when concentrated hydrochloric acid is added to a solution of copper(II) sulphate.

Method
Recall the effect of ligand substitution on shape, co-ordination number, and colour in a complex ion.

Answer
The copper(II)-aqua ion is *octahedral*. When small water molecules are replaced by bigger chloride ions, a *tetrahedral* complex is likely to be formed. The *co-ordination number* will fall from 6 to 4. There will probably be a change of colour in the solution.

> **Comment**
> The blue solution of copper(II) ions turns yellow-green once an excess of hydrochloric acid has been added and the tetrachlorocuprate(II) ion is formed:
>
> | *Equation:* | $[Cu(H_2O)_6]^{2+} + 4Cl^- \rightleftharpoons [CuCl_4]^{2-} + 6H_2O$ | |
> |---|---|---|
> | *Colour:* | blue | yellow-green |
> | *Shape:* | octahedral | tetrahedral |
> | *Co-ordination number:* | 6 | 4 |
>
> When the solution is diluted, the decreased concentration of chloride ions forces the equilibrium back to the hexaaqua ion and the solution turns blue again.

Substitution by bidentate and multidentate ligands

When a ligand with two donor atoms attacks a metal-aqua ion, two water molecules are replaced in the first instance. For example, a bidentate ligand (see section 14.4.1) such as ethane-1,2-diamine (*ethylenediamine* or *en*) reacts with a metal(II)-aqua ion according to the equation:

$$[M(H_2O)_6]^{2+} + H_2NCH_2CH_2NH_2 \rightleftharpoons [M(H_2O)_4(H_2NCH_2CH_2NH_2)]^{2+} + 2H_2O$$

As more ligand is added, further substitution can occur until all the water molecules have been replaced:

$$[M(H_2O)_4(H_2NCH_2CH_2NH_2)]^{2+} + H_2NCH_2CH_2NH_2 \rightleftharpoons [M(H_2O)_2(H_2NCH_2CH_2NH_2)_2]^{2+} + 2H_2O$$

$$[M(H_2O)_2(H_2NCH_2CH_2NH_2)_2]^{2+} + H_2NCH_2CH_2NH_2 \rightleftharpoons [M(H_2NCH_2CH_2NH_2)_3]^{2+} + 2H_2O$$

In these reactions with ethylenediamine, the equilibrium position lies well over to the right-hand side; the equilibrium constants for the formation of tris(ethylenediamine) complexes are of the order of 10^{20}. This large value means that the resulting complexes are much more stable than the aqua ions from which they were formed. Note that, because the donor atoms are small, there is no change in shape and the complexes remain octahedral.

Metal(III)-aqua ions react similarly so that the equation for the reaction with an excess of ethane-1,2-diamine is:

$$[M(H_2O)_6]^{3+} + 3H_2NCH_2CH_2NH_2 \rightleftharpoons [M(H_2NCH_2CH_2NH_2)_3]^{3+} + 6H_2O$$

These metal(III)-tris(ethylenediamine) complexes are even more stable than the metal(II) complexes; the equilibrium constant for the above reaction is of the order of 10^{30}.

This extra stability is known as the **chelate effect**. Complexes in which bidentate or multidentate ligands bond to one metal ion only are known as chelates. The ligand forms a five- or six-membered ring with the metal ion, as shown in Fig 17.

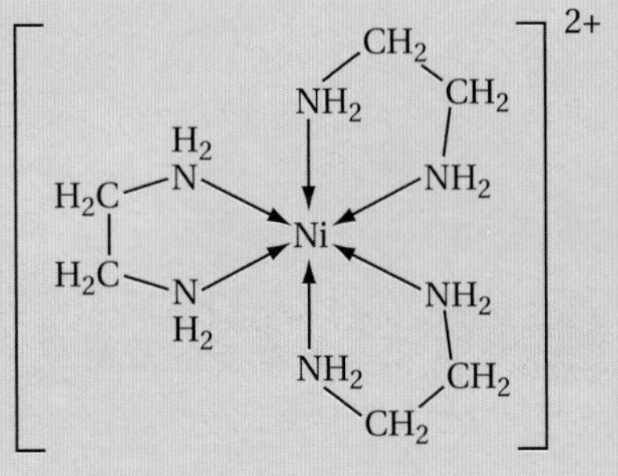

Fig 17
Chelating ethylenediamine.

Chelating ligands can be present in complexes which also contain unidentate ligands as in, for example, the ion shown in Fig 18, *trans*-dichlorobis(ethylenediamine)cobalt(III).

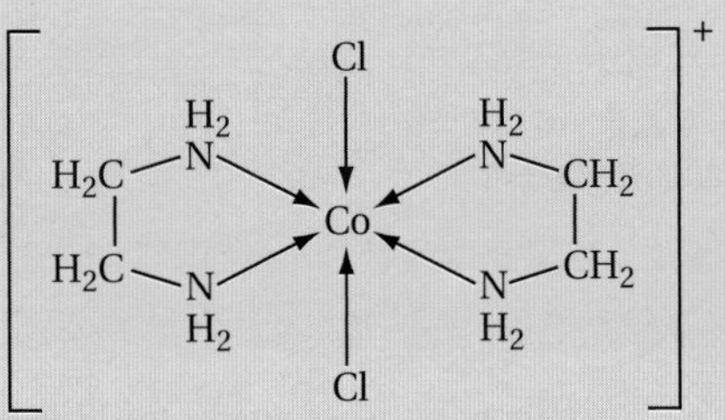

E Hint: When drawing complex ions with *multidentate ligands*, *first* draw the six co-ordinate bonds arranged octahedrally around the central metal atom, *then* fit the ligands to these bonds.

Fig 18
Unidentate chloride ions as part of a bidentate chelate complex.

Another common bidentate ligand is the di-anion ethanedioate, $C_2O_4^{2-}$. This ligand forms a very stable complex with iron(III) ions, $[Fe(C_2O_4)_3]^{3-}$, shown in Fig 19.

Fig 19
The $[Fe(C_2O_4)_3]^{3-}$ ion.

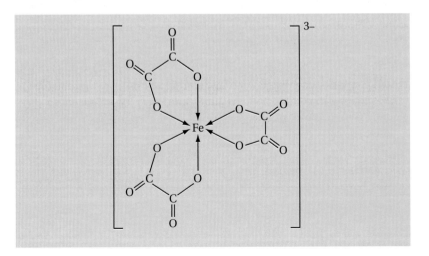

Multidentate ligands form even more stable complexes than do bidentate ligands. Haem in blood (see sections 14.4.2 and 14.4.7) is an example taken from iron chemistry. The *hexadentate* ligand EDTA^{4-} reacts with metal(II)-aqua ions to form a complex in which the ligand occupies all six octahedral sites around the metal ion, liberating six water molecules:

$$[M(H_2O)_6]^{2+} + EDTA^{4-} \rightleftharpoons [M(EDTA)]^{2-} + 6H_2O$$

The EDTA^{4-} ligand (see Fig 10, section 14.4.2) finds many uses in analytical and industrial chemistry. The complexes formed are very stable, so that all the equilibria are completely over to the right-hand side and very few aqua ions are present. Reactions involving the metal-aqua ion are therefore not possible in the presence of an excess of EDTA^{4-}, so that metal ions can be kept in solution (**sequestered**), even when anions which normally cause precipitation, such as OH$^-$ or CO$_3^{2-}$, are added.

Thus, addition of an excess of the sodium salt of EDTA^{4-} to aqueous copper(II) sulphate, followed by aqueous sodium hydroxide, does not give any precipitate of copper(II) hydroxide because the Cu^{2+} ions have been *sequestered*.

There is a thermodynamic reason for the stability of metal chelates. Consider the equation for the reaction of EDTA^{4-} given opposite. The left-hand side of the equation has *two* particles, $[M(H_2O)_6]^{2+}$ and EDTA^{4-}, whereas the right-hand side of this equation has *seven* particles, $[M(EDTA)]^{2-}$ and six H$_2$O molecules. This considerable increase in randomness results in an increase in *entropy* (see section 14.1.2).

Recall that a chemical reaction becomes *feasible* if the change in free energy, $\Delta G^{\ominus}$, is negative or zero. The equation

$$\Delta G^{\ominus} = \Delta H^{\ominus} - T\Delta S^{\ominus}$$

shows that there will *always* be a negative free-energy change if the enthalpy change, $\Delta H^{\ominus}$, is *negative* and the entropy change, $\Delta S^{\ominus}$, is *positive*. Even when $\Delta H^{\ominus}$ is *positive* (but not very much so), it will readily be outweighed by the $T\Delta S^{\ominus}$ term if the entropy change is *large and positive*.

In the formation of a chelate, $\Delta S^{\ominus}$ is indeed *large and positive* (two particles form seven particles). $\Delta H^{\ominus}$ for such reactions is usually quite small because the bonds formed are the same in number and rather similar in strength to the bonds broken. Even if $\Delta H^{\ominus}$ were to be slightly positive, this term would be outweighed heavily by $-T\Delta S^{\ominus}$. Consequently, $\Delta G^{\ominus}$ is *always very negative* and the reaction is *always feasible*. Reactions of this kind are sometimes described as being *entropy driven* because the enthalpy term can usually be ignored.

Use of this sequestering ability is made in water softening, where the deposition of calcium carbonate in pipes can be prevented.

The increased entropy leads to a large and positive $T\Delta S^{\ominus}$ term which, because it appears in the equation with a negative sign, contributes significantly to ensuring that $\Delta G^{\ominus}$ is very negative.

A2 5 Sample module test

1 The table below lists a number of mean bond enthalpy values.

Bond	Mean bond enthalpy/kJ mol^{-1}
C–C	348
C=C	612
C–H	413
O–H	463

(a) Explain the meaning of the term *mean bond enthalpy*.

....

....

....

(3 marks)

(b) Given that the enthalpy of combustion to form carbon dioxide and steam is
–2102 kJ mol^{-1} for propane and –1977 kJ mol^{-1} for propene, determine the enthalpy
change for the oxidation of 1 mol of propane to propene and steam

$$C_3H_8(g) + \tfrac{1}{2}O_2(g) \rightarrow C_3H_6(g) + H_2O(g)$$

using equations or a cycle to support your answer.

....

....

....

....

....

(3 marks)

(c) State the number and type of bonds broken and formed in the oxidation of propane to propene and steam. Use the mean bond enthalpies in the table, together with your answer to part (b), to calculate the bond enthalpy of the O=O bond in the oxygen molecule.

Bonds broken..

....

....

Bonds formed ....

....

....

Bond enthalpy of O=O ....

....

....

....

....

(4 marks)

$\overline{10}$

2 Sulphur dioxide reacts with oxygen to form sulphur trioxide according to the equation

$$2SO_2(g) + O_2(g) \rightleftharpoons 2SO_3(g)$$

Data for this reaction are shown in the table below.

	$\Delta H_f^\ominus$/kJ mol^{-1}	$S^\ominus$/J K^{-1} mol^{-1}
$SO_3(g)$	−396	+257
$SO_2(g)$	−297	+248
$O_2(g)$	0	+204

(a)   Determine the standard enthalpy, the standard entropy and the standard free-energy changes at 298 K for this reaction.

$\Delta H_{298}^\ominus$ ....

....

....

....

....

$\Delta S_{298}^\ominus$ ....

....

....

....

....

$\Delta G_{298}^\ominus$ ....

....

....

....

....

(7 marks)

(b) The reaction is said to be feasible. In terms of free-energy change, explain the meaning of the term *feasible*. Calculate the temperature at which the reaction between sulphur dioxide and oxygen ceases to be feasible.

Feasible reaction ....

....

Temperature ....

....

....

(3 marks)

$\overline{10}$

3 (a) Write equations to show what happens when each of the following chlorides is added to water, and predict approximate values for the pH of the resulting solutions.

(i)   sodium chloride

Equation ....

....

pH ....

....

(ii)   silicon tetrachloride

Equation ....

....

pH ....

....

(4 marks)

(b)   What is the relationship between bond type in the chlorides of Period 3 elements and the pH of the solutions which result from addition of these chlorides to water?

....

....

....

(2 marks)

(c) Write equations to show what happens when each of the following is heated in an excess of oxygen and, in each case, state the type of bonding found in the product.

(i)   sodium

Equation ....

....

Type of bonding in product ....

....

(ii) phosphorus

Equation ....

....

Type of bonding in product ....

....

(4 marks)

$\overline{10}$

4 Use the data below to answer the questions that follow.

Reaction at 298 K	$E^{\ominus}$/ V
$Ag^+(aq) \; + \; e^- \; \rightarrow \; Ag(s)$	+0.80
$AgF(s) \quad + \; e^- \; \rightarrow \; Ag(s) + F^-(aq)$	+0.78
$AgCl(s) \; + \; e^- \; \rightarrow \; Ag(s) + Cl^-(aq)$	+0.22
$AgBr(s) \; + \; e^- \; \rightarrow \; Ag(s) + Br^-(aq)$	+0.07
$H^+(aq) \quad + \; e^- \; \rightarrow \; \frac{1}{2}H_2(g)$	0.00
$D^+(aq) \quad + \; e^- \; \rightarrow \; \frac{1}{2}D_2(g)$	−0.004
$AgI(s) \quad + \; e^- \; \rightarrow \; Ag(s) + I^-(aq)$	−0.15

The symbol D denotes deuterium, which is heavy hydrogen, ^2_1H.

(a) By considering electron transfer, state what is meant by the term *oxidising agent*.

....

(1 mark)

(b) State which of the two ions, $H^+(aq)$ or $D^+(aq)$, is the more powerful oxidising agent. Write an equation for the spontaneous reaction which occurs when a mixture of aqueous H^+ ions and D^+ ions are in contact with a mixture of hydrogen and deuterium gases. Deduce the e.m.f. of the cell in which this reaction would occur spontaneously.

Stronger oxidising agent ....

Equation ....

....

....

....

e.m.f. ....

....

(3 marks)

(c) Write an equation for the spontaneous reaction which occurs when aqueous F^- ions and Cl^- ions are in contact with a mixture of solid AgF and solid AgCl. Deduce the e.m.f. of the cell in which this reaction would occur spontaneously.

Equation ....

....

....

e.m.f. ....

....

(2 marks)

(d) Silver does not usually react with dilute solutions of strong acids to liberate hydrogen.

   (i)   State why this is so.

       ....

   (ii) Suggest a hydrogen halide which might react with silver to liberate hydrogen in aqueous solution. Write an equation for the reaction and deduce the e.m.f. of the cell in which this reaction would occur spontaneously.

       Hydrogen halide ....

       Equation ....

       e.m.f. ....

(4 marks)

$\overline{10}$

5 Vanadium(IV) chloride, VCl_4, is a brown liquid which behaves as a Lewis acid. It dissolves in hexane without reaction, but reacts violently with water to form a blue solution of the oxovanadium(IV) ion according to the equation:

$$VCl_4(l) + 3H_2O(l) \rightarrow VO^{2+}(aq) + 2H_3O^+(aq) + 4Cl^-(aq)$$

(a) Deduce the type of bonding present in VCl_4.

....

(1 mark)

(b) Suggest why VCl_4 reacts with water but not with hexane.

....

....

(2 marks)

(c) In order to determine the percentage of vanadium in a given sample of VCl_4, a weighed sample was added to an excess of water. The chloride ions were removed by precipitation and filtration, and the solution of vanadium(IV) was titrated with standard potassium manganate(VII).

A 0.475 g sample of VCl_4 required 24.1 cm^3 of a solution of 0.0200 mol dm^{-3} $KMnO_4$ for complete reaction. In the titration, the reaction occurring is the oxidation of vanadium(IV) to vanadium(V).

(i) Explain why it is necessary to remove the chloride ions from the vanadium solution before the titration with potassium manganate(VII).

....

....

(ii) Calculate the percentage by mass of vanadium in the sample analysed above.

....

....

....

(7 marks)

$\overline{10}$

6 Study the passage below and answer the questions which follow.

Crystalline iron(III) nitrate nonahydrate, $Fe(NO_3)_3.9H_2O$, has a very pale violet colour and contains the ion $[Fe(H_2O)_6]^{3+}$. When added to water, the crystals dissolve to form a brown solution. Treatment of this brown solution with concentrated nitric acid yields a very pale violet solution.

(a) Name the shape of the $[Fe(H_2O)_6]^{3+}$ ion.

....

(1 mark)

(b) Write an equation to show the $[Fe(H_2O)_6]^{3+}$ ion behaving as an acid in aqueous solution.

....

(1 mark)

(c) Deduce the formula of the species responsible for the brown colour of the solution described above.

....

(1 mark)

(d) Explain why the addition of concentrated nitric acid causes the colour of the solution to change from brown to very pale violet.

....

....

....

(2 marks)

(e) When concentrated hydrochloric acid is added to the brown solution of iron(III) nitrate, however, a yellow solution containing $[FeCl_4]^-$ ions is formed. Give two reasons for a colour change in this reaction.

Reason 1 ....

Reason 2 ....

(2 marks)

(f) When an excess of magnesium metal is added to an aqueous solution of iron(III) nitrate, effervescence occurs and a brown precipitate forms. Identify the gas evolved, give the formula of the brown precipitate and construct an equation, or equations, for the reaction occurring.

Identity of gas ....

Formula of brown precipitate ....

Equation(s) ....

....

....

(3 marks)

$\overline{10}$

7 (a) Explain what is meant by a substitution reaction of a transition-metal aqua ion. Discuss, as examples, the reactions of copper(II) and cobalt(II) aqua ions with ammonia and with chloride ions.

In your answer you should state the shape and colour of each complex formed, write equations for all the reactions occurring and account briefly for any differences in the shapes of the complexes formed.

(15 marks)

(b) A proprietary moss killer consists of a mixture of sand and hydrated iron(II) sulphate. Outline the plan of an experiment to determine the percentage by mass of iron(II) in a sample of this moss killer, using a standard solution of potassium dichromate(VI). You must show in your answer how you would calculate the result of the determination.

(8 marks)

(c) Explain the following observations.

(i) Sulphur dioxide reacts quickly with oxygen when vanadium(V) oxide is present.

(ii) When a solution of potassium manganate(VII) is added dropwise to an acidified solution of sodium ethanedioate, the purple colour is decolourised only slowly at first but then more rapidly as more potassium manganate(VII) is added.

(7 marks)

....

....

....

....

....

....

....

....

....

....

Answers

1 (a) energy or heat
to break a covalent bond
averaged over several compounds   3

(b) $C_3H_8(g) + 5O_2(g) \rightarrow 3CO_2(g) + 4H_2O(g)$
$C_3H_6(g) + 4\frac{1}{2}O_2(g) \rightarrow 3CO_2(g) + 3H_2O(g)$
_____
$C_3H_8(g) + \frac{1}{2}O_2(g) \rightarrow C_3H_6(g) + H_2O(g)$
$\Delta H = -2102 + 1977$
$= -125 \text{ kJ mol}^{-1}$ 3

(c) Bonds broken:
$2 \times (C–C) + 8 \times (C–H) + \frac{1}{2}(O=O)$

Bonds formed:
$1 \times (C–C) + 6 \times (C–H) + 1 \times (C=C) + 2 \times (O–H)$

Bond enthalpy of O=O:
$\Delta H = \Sigma B(\text{bonds broken}) - \Sigma B(\text{bonds formed})$
$\therefore -125 = B(C–C) + 2B(C–H) + \frac{1}{2}B(O=O) - B(C=C)$
$\quad\quad - 2B(O–H)$
$\therefore B(O=O) = 2(-125 - 348 - 2 \times 413 + 612 + 2 \times 463)$
$= 478 \text{ kJ mol}^{-1}$ 4

2 (a) $\Delta H^{\ominus} = \Sigma \Delta H_f^{\ominus}(\text{products}) - \Sigma \Delta H_f^{\ominus}(\text{reactants})$
$= 2 \times (-396) - 2 \times (-297)$
$= -198 \text{ kJ mol}^{-1}$

$\Delta S^{\ominus} = \Sigma S^{\ominus}(\text{products}) - \Sigma S^{\ominus}(\text{reactants})$
$= (2 \times 257) - 204 - (2 \times 248)$
$= -186 \text{ J K}^{-1} \text{ mol}^{-1}$

$\Delta G^{\ominus} = \Delta H^{\ominus} - T\Delta S^{\ominus}$
$= (-198) - \dfrac{298 \times (-186)}{1000}$
$= -143 \text{ kJ mol}^{-1}$ 7

(b) one for which $\Delta G \leq 0$
$T = \Delta H / \Delta S$ when $\Delta G = 0$
$= \dfrac{(-198) \times 1000}{-186}$
$= 1065 \text{ K}$ 3

3 (a) $NaCl + H_2O \rightarrow Na^+(aq) + Cl^-(aq)$
pH = 7
$SiCl_4 + 4H_2O \rightarrow Si(OH)_4 + 4HCl$
pH = 0   4

(b) ionic chlorides → neutral solutions
covalent chlorides → acidic solutions   2

(c) $2Na + \frac{1}{2}O_2 \rightarrow Na_2O$
ionic

$P_4 + 5O_2 \rightarrow P_4O_{10}$
covalent   4

4 (a) electron acceptor   1

(b) H^+
$H^+(aq) + e^- \rightarrow \frac{1}{2}H_2(g)$
$D^+(aq) + e^- \rightarrow \frac{1}{2}D_2(g)$
_____
$H^+(aq)_{\frac{1}{2}} + {}^1D_2(g) \rightarrow D^+(aq) + {}^1H_2(g)$
$0.00 - (-0.004) = +0.004 \text{ V}$ 3

(c) $AgF(s) + Cl^-(aq) \rightarrow AgCl(s) + F^-(aq)$
e.m.f. = $+0.78 - 0.22 = +0.56 \text{ V}$ 2

(d) silver lies above hydrogen in the electrochemical series
HI
$H^+(aq) + I^-(aq) + Ag(s) \rightarrow AgI(s) + \frac{1}{2}H_2(g)$
$0.00 - (-0.15) = +0.15 \text{ V}$ 4

5 (a) covalent   1

(b) water is a Lewis base (or has lone pairs)
hexane not a Lewis base (or no lone pairs)   2

(c) Cl^- reacts with MnO_4^-
giving inaccurate result   2
moles $MnO_4^- = \dfrac{24.1 \times 0.0200}{1000} = 4.82 \times 10^{-4}$
moles V $= 5 \times 4.82 \times 10^{-4} = 2.41 \times 10^{-3}$
mass of V $= 51 \times 2.41 \times 10^{-3} = 1.23 \times 10^{-1}\text{g}$
% V $= \dfrac{1.23 \times 10^{-1} \times 100}{0.475}$
$= 25.9 \text{ \% by mass}$ 5

6 (a) octahedral   1

(b) $[Fe(H_2O)_6]^{3+} + H_2O \rightarrow [Fe(OH)(H_2O)_5]^{2+} + H_3O^+$ 1

(c) $[Fe(OH)(H_2O)_5]^{2+}$ 1

(d) HNO_3 addition $\equiv H_3O^+$ addition
reverses equilibrium *or* forms $[Fe(H_2O)_6]^{3+}$ 2

(e) change of ligand
change of shape *or* co-ordination number   2

(f) H_2
$[Fe(OH)_3(H_2O)_3]$
$3Mg + 2[Fe(H_2O)_6]^{3+}$
$\rightarrow 3Mg^{2+} + 3H_2 + 2[Fe(OH)_3(H_2O)_3]$ 3

7 (a) replacement of water *or* M–O bond breaks
by another ligand *or* another ligand replaces

$$[Cu(H_2O)_6]^{2+} + 4NH_3 \rightarrow [Cu(NH_3)_4(H_2O)_2]^{2+} + 4H_2O$$
octahedral
deep blue

$$[Co(H_2O)_6]^{2+} + 6NH_3 \rightarrow [Co(NH_3)_6]^{2+} + 6H_2O$$
octahedral
pale brown

$$[Cu(H_2O)_6]^{2+} + 4Cl^- \rightarrow [CuCl_4]^{2-} + 6H_2O$$
tetrahedral
yellow-green

$$[Co(H_2O)_6]^{2+} + 4Cl^- \rightarrow [CoCl_4]^{2-} + 6H_2O$$
tetrahedral
blue

Cl^- larger than H_2O 15

(b) known mass
extract with H_2O
add dil. H_2SO_4
titrate with $Cr_2O_7^{2-}$
add indicator
$$6Fe^{2+} + Cr_2O_7^{2-} + 14H^+ \rightarrow 6Fe^{3+} + 2Cr^{3+} + 7H_2O$$
moles $Fe^{2+} = 6 \times$ moles $Cr_2O_7^{2-}$
g $Fe^{2+} =$ moles $Fe^{2+} \times A_r$
% $Fe^{2+} = ($mass $Fe^{2+}/$initial mass$) \times 100$ 8

(c) V_2O_5 is a catalyst
finds alternative route *or* lowers E_A
or V changes oxidation state
$$V_2O_5 + SO_2 \rightarrow V_2O_4 + SO_3$$
$$V_2O_4 + \tfrac{1}{2}O_2 \rightarrow V_2O_5$$
Mn^{2+} is a catalyst
none present initially $\therefore$ slow
or two negative ions reacting
more present as reaction proceeds $\therefore$ faster
autocatalysis   7

The Periodic Table of the Elements

Key

| relative atomic mass | 6.9 **Li** Lithium |
| atomic number | 3 |

I	II												III	IV	V	VI	VII	0
																		4.0 **He** Helium 2
6.9 **Li** Lithium 3	9.0 **Be** Beryllium 4												10.8 **B** Boron 5	12.0 **C** Carbon 6	14.0 **N** Nitrogen 7	16.0 **O** Oxygen 8	19.0 **F** Fluorine 9	20.2 **Ne** Neon 10
23.0 **Na** Sodium 11	24.3 **Mg** Magnesium 12												27.0 **Al** Aluminium 13	28.1 **Si** Silicon 14	31.0 **P** Phosphorus 15	32.1 **S** Sulphur 16	35.5 **Cl** Chlorine 17	39.9 **Ar** Argon 18
39.1 **K** Potassium 19	40.1 **Ca** Calcium 20	45.0 **Sc** Scandium 21	47.9 **Ti** Titanium 22	50.9 **V** Vanadium 23	52.0 **Cr** Chromium 24	54.9 **Mn** Manganese 25	55.8 **Fe** Iron 26	58.9 **Co** Cobalt 27	58.7 **Ni** Nickel 28	63.5 **Cu** Copper 29	65.4 **Zn** Zinc 30		69.7 **Ga** Gallium 31	72.6 **Ge** Germanium 32	74.9 **As** Arsenic 33	79.0 **Se** Selenium 34	79.9 **Br** Bromine 35	83.8 **Kr** Krypton 36
85.5 **Rb** Rubidium 37	87.6 **Sr** Strontium 38	88.9 **Y** Yttrium 39	91.2 **Zr** Zirconium 40	92.9 **Nb** Niobium 41	95.9 **Mo** Molybdenum 42	98.9 **Tc** Technetium 43	101.1 **Ru** Ruthenium 44	102.9 **Rh** Rhodium 45	106.4 **Pd** Palladium 46	107.9 **Ag** Silver 47	112.4 **Cd** Cadmium 48		114.8 **In** Indium 49	118.7 **Sn** Tin 50	121.8 **Sb** Antimony 51	127.6 **Te** Tellurium 52	126.9 **I** Iodine 53	131.3 **Xe** Xenon 54
132.9 **Cs** Caesium 55	137.3 **Ba** Barium 56	138.9 **La** Lanthanum 57 *	178.5 **Hf** Hafnium 72	180.9 **Ta** Tantalum 73	183.8 **W** Tungsten 74	186.2 **Re** Rhenium 75	190.2 **Os** Osmium 76	192.2 **Ir** Iridium 77	195.1 **Pt** Platinum 78	197.0 **Au** Gold 79	200.6 **Hg** Mercury 80		204.4 **Tl** Thallium 81	207.2 **Pb** Lead 82	209.0 **Bi** Bismuth 83	(209) **Po** Polonium 84	(210) **At** Astatine 85	(222) **Rn** Radon 86
(223) **Fr** Francium 87	(226) **Ra** Radium 88	(227) **Ac** Actinium 89 †																

* 58–71 Lanthanides

† 90–103 Actinides

140.1 **Ce** Cerium 58	140.9 **Pr** Praseodymium 59	144.2 **Nd** Neodymium 60	(145) **Pm** Promethium 61	150.4 **Sm** Samarium 62	152.0 **Eu** Europium 63	157.3 **Gd** Gadolinium 64	158.9 **Tb** Terbium 65	162.5 **Dy** Dysprosium 66	164.9 **Ho** Holmium 67	167.3 **Er** Erbium 68	168.9 **Tm** Thulium 69	173.0 **Yb** Ytterbium 70	175.0 **Lu** Lutetium 71
232.0 **Th** Thorium 90	231.0 **Pa** Protactinium 91	238.0 **U** Uranium 92	(237) **Np** Neptunium 93	(244) **Pu** Plutonium 94	(243) **Am** Americium 95	(247) **Cm** Curium 96	(247) **Bk** Berkelium 97	(251) **Cf** Californium 98	(252) **Es** Einsteinium 99	(257) **Fm** Fermium 100	(258) **Md** Mendelevium 101	(259) **No** Nobelium 102	(262) **Lr** Lawrencium 103

Collins Support Materials for AQA – ORDER FORM

This booklet covers one module from the AQA chemistry course at A-level.
If you would like to order further copies from the series, please send a completed copy of this page to Collins by telephone, fax or post.

Title	ISBN	Price	Evaluation copy	Order quantity
1 Atomic Structure, Bonding and Periodicity	000327701 1	£4.25		
2 Foundation Physical and Inorganic Chemistry	000327702 X	£4.25		
3 Introduction to Organic Chemistry	000327703 8	£4.25		
4 Further Physical and Organic Chemistry	000327704 6	£5.99		
5 Thermodynamics and Further Inorganic Chemistry	000327705 4	£5.99		
TOTAL ORDER VALUE				

Also available:

Collins Advanced Modular Sciences – Chemistry
comprehensive textbooks to support the AQA specification

Title	ISBN	Price	Evaluation copy	Order quantity
Chemistry AS	000327753 4	£17.99		
Chemistry A2	000327754 2	£17.99		
TOTAL ORDER VALUE				

Details of other A-level titles in this series are available on our website:

www.**Collins**Education.com

Please fill in your details and send your order to the address below:

Name Address	Tel: 0870 0100 442 Fax: 0141 306 3750 Post: Collins Educational HarperCollins Publishers FREEPOST GW2446 GLASGOW G64 1BR